INTERNATIONAL SERIES OF MONOGRAPHS ON PHYSICS

International Series of Monographs on Physics

148. H. Asada, T. Futamase, P. Hogan: *Equations of motion in general relativity*
147. A. Yaouanc, P. Dalmas de Réotier: *Muon spin rotation, relaxation, and resonance*
146. B. McCoy: *Advanced statistical mechanics*
145. M. Bordag, G.L. Klimchitskaya, U. Mohideen, V.M. Mostepanenko: *Advances in the Casimir effect*
144. T.R. Field: *Electromagnetic scattering from random media*
143. W. Götze: *Complex dynamics of glass-forming liquids–a mode-coupling theory*
142. V.M. Agranovich: *Excitations in organic solids*
141. W.T. Grandy: *Entropy and the time evolution of macroscopic systems*
140. M. Alcubierre: *Introduction to* $3+1$ *numerical relativity*
139. A. L. Ivanov, S. G. Tikhodeev: *Problems of condensed matter physics–quantum coherence phenomena in electron-hole and coupled matter-light systems*
138. I. M. Vardavas, F. W. Taylor: *Radiation and climate*
137. A. F. Borghesani: *Ions and electrons in liquid helium*
136. C. Kiefer: *Quantum gravity*, second edition
135. V. Fortov, I. Iakubov, A. Khrapak: *Physics of strongly coupled plasma*
134. G. Fredrickson: *The equilibrium theory of inhomogeneous polymers*
133. H. Suhl: *Relaxation processes in micromagnetics*
132. J. Terning: *Modern supersymmetry*
131. M. Mariño: *Chern–Simons theory, matrix models, and topological strings*
130. V. Gantmakher: *Electrons and disorder in solids*
129. W. Barford: *Electronic and optical properties of conjugated polymers*
128. R. E. Raab, O. L. de Lange: *Multipole theory in electromagnetism*
127. A. Larkin, A. Varlamov: *Theory of fluctuations in superconductors*
126. P. Goldbart, N. Goldenfeld, D. Sherrington: *Stealing the gold*
125. S. Atzeni, J. Meyer-ter-Vehn: *The physics of inertial fusion*
123. T. Fujimoto: *Plasma spectroscopy*
122. K. Fujikawa, H. Suzuki: *Path integrals and quantum anomalies*
121. T. Giamarchi: *Quantum physics in one dimension*
120. M. Warner, E. Terentjev: *Liquid crystal elastomers*
119. L. Jacak, P. Sitko, K. Wieczorek, A. Wojs: *Quantum Hall systems*
118. J. Wesson: *Tokamaks*, Third edition
117. G. Volovik: *The Universe in a helium droplet*
116. L. Pitaevskii, S. Stringari: *Bose–Einstein condensation*
115. G. Dissertori, I.G. Knowles, M. Schmelling: *Quantum chromodynamics*
114. B. DeWitt: *The global approach to quantum field theory*
113. J. Zinn-Justin: *Quantum field theory and critical phenomena*, Fourth edition
112. R.M. Mazo: *Brownian motion–fluctuations, dynamics, and applications*
111. H. Nishimori: *Statistical physics of spin glasses and information processing–an introduction*
110. N.B. Kopnin: *Theory of nonequilibrium superconductivity*
109. A. Aharoni: *Introduction to the theory of ferromagnetism*, Second edition
108. R. Dobbs: *Helium three*
107. R. Wigmans: *Calorimetry*
106. J. Kübler: *Theory of itinerant electron magnetism*
105. Y. Kuramoto, Y. Kitaoka: *Dynamics of heavy electrons*
104. D. Bardin, G. Passarino: *The standard model in the making*
103. G.C. Branco, L. Lavoura, J.P. Silva: *CP Violation*
102. T.C. Choy: *Effective medium theory*
101. H. Araki: *Mathematical theory of quantum fields*
100. L. M. Pismen: *Vortices in nonlinear fields*
99. L. Mestel: *Stellar magnetism*
98. K. H. Bennemann: *Nonlinear optics in metals*
94. S. Chikazumi: *Physics of ferromagnetism*
91. R. A. Bertlmann: *Anomalies in quantum field theory*
90. P. K. Gosh: *Ion traps*
87. P. S. Joshi: *Global aspects in gravitation and cosmology*
86. E. R. Pike, S. Sarkar: *The quantum theory of radiation*
83. P. G. de Gennes, J. Prost: *The physics of liquid crystals*
73. M. Doi, S. F. Edwards: *The theory of polymer dynamics*
69. S. Chandrasekhar: *The mathematical theory of black holes*
51. C. Møller: *The theory of relativity*
46. H. E. Stanley: *Introduction to phase transitions and critical phenomena*
32. A. Abragam: *Principles of nuclear magnetism*
27. P. A. M. Dirac: *Principles of quantum mechanics*
23. R. E. Peierls: *Quantum theory of solids*

Equations of Motion in General Relativity

H. Asada
Hirosaki University

T. Futamase
Tohoku University

P. A. Hogan
University College Dublin

OXFORD
UNIVERSITY PRESS

OXFORD
UNIVERSITY PRESS

Great Clarendon Street, Oxford, OX2 6DP,
United Kingdom

Oxford University Press is a department of the University of Oxford.
It furthers the University's objective of excellence in research, scholarship,
and education by publishing worldwide. Oxford is a registered trade mark of
Oxford University Press in the UK and in certain other countries

First published 2011

Published in the United States of America by Oxford University Press
198 Madison Avenue, New York, NY 10016, United States of America

British Library Cataloguing in Publication Data
Data available

ISBN 978-0-19-958410-9

Preface

In this book we focus attention on two aspects of equations of motion in general relativity: the motion of extended bodies and the motion of small black holes, both charged and uncharged, in external gravitational and electromagnetic fields. Our objective is to offer a guide to prospective researchers into these areas of the subject and to point out open questions and topics that are ripe for further development. The book is aimed at first year postgraduate students or final year undergraduates who have already taken a good introductory course in general relativity, with the necessary differential geometry. Although the presentation is strongly theoretical, some aspects may be useful to a broader range of gravitational physicists.

It is well known that the equations of general relativity take the form of Newton's equations in an appropriate limit. This is the so-called Newtonian limit and it forms the basis for developing the post-Newtonian approximation which is of central importance in discussing equations of motion. This limit is quite delicate in that it is not uniform in time since, for example, a binary system exhibits inspiral due to gravitational radiation reaction with consequent domination by gravitational waves over Newtonian effects. If proper account is not taken of this the mathematics will become meaningless with the appearance of divergent integrals.

In derivations of equations of motion of compact objects in general relativity singular functions are sometimes used which lead to divergent quantities. The theory presented in this book leads to equations of motion in the post-Newtonian approximation in which the strong internal gravity of the masses is taken into account. To achieve this we describe in detail the method of the point particle limit in which the equations of motion of extended objects are expressed in terms of surface integrals. The behavior of the surface integrals is carefully elucidated. This method is successful up to the third post-Newtonian order of approximation.

Aside from the solar system a principal astronomical system utilized for the experimental verification of Einstein's theory of gravity is a binary star system. The equations of motion for two extended bodies moving under their mutual gravitational attraction, whose derivation is described here, enable theoretical predictions to be calculated for the advance of the periastron, light deflection, Shapiro time delay, orbital period decay due to energy loss via gravitational radiation and spin precession. All of these phenomena are treated in this book.

Our presentation of equations of motion of small black holes is based upon recent research in which the approximate equations of motion emerge from the field equations and the requirement that the wave fronts of radiation produced by the motion of the small black hole are smoothly deformed spheres near the black hole. The black hole is introduced as a small perturbation of a background space-time which is a model of the external field in which the black hole moves, so that this space-time is itself a solution of appropriate field equations. The perturbation is singular on a time-like world-line of the background space-time and the equations of motion are differential equations for this world-line. Thus in this theory the field of a small Reissner–Nordström black hole, for example, is modeled as a small perturbation of a flat background space-time because there is no external field. The world-line of the black hole is in this case a time-like geodesic of the flat background. Thus the charged black hole does not "run away" in the manner of a Dirac charged mass and nor do its equations of motion have an infinite self-energy term requiring elimination by a "renormalization" of the mass, which is difficult to justify. Our presentation serves to demonstrate how to construct more general models of small black holes moving relativistically in external fields in a way that opens up a whole range of interesting and difficult theoretical problems.

The book ends with a discussion of few-body systems in Newtonian gravity and in general relativity. In particular we focus on so–called choreographic configurations in which each body in the system moves in a single closed orbit. We describe connections between the choreographic solutions and the gravitational waves predicted by general relativity. General relativistic effects cause a complicated flower-like pattern to appear in the orbits of masses in binary systems. Remarkably it is possible to have a three-body system with a figure-eight orbit in general relativity, if dissipative effects are ignored, just like in Newtonian theory.

Acknowledgements

We thank the Canon Foundation in Europe for making it possible for us to initiate our collaboration and we are grateful to the Japan Society for the Promotion of Science (JSPS) and Tohoku University for supporting our collaboration over many years.

Contents

1
Introduction

1.1 Equations of motion of extended bodies

The specialized field of equations of motion within Einstein's general theory of relativity is, in effect, general relativistic celestial mechanics. Its original motivation was the theoretical insight that since Einstein's field equations are nonlinear they must somehow incorporate, in appropriate physical situations such as n-body self-gravitating systems, the motion of the sources of the gravitational fields they are used to model. This insight was first confirmed by the pioneering work of Einstein, Infeld and Hoffmann (1938). They calculated, using the vacuum field equations, approximate equations of motion for an n-body system in which the sources were viewed as points. The corresponding derivation of the n-body equations of motion for extended bodies, in which the internal structure of the bodies was taken into account, was first performed by Fock (1939) and Papapetrou (1951*a*). Approximation schemes are necessary for this work where the basic assumptions are that the gravitational fields are weak, the motions are slow compared to the speed of light and, in the latter cases in which the extent of the gravitating bodies is taken into account, the equations of motion obtained apply when the bodies are a large distance apart compared to their dimensions. This early development of equations of motion is described in the texts by Bergmann (1947), Infeld and Plebanski (1960), Fock (1959), and Levi-Civita (1965). The approximation schemes, referred to generally as post-Newtonian approximation schemes since they are designed to evaluate deviations in general relativity from Newtonian gravitation theory predictions, underwent considerable refinements and applications in the work of Chandrasekhar (1965, 1969) [see also Chandrasekhar and Nutku (1969) and Chandrasekhar and Esposito (1970)] and of Synge (1970*a*) [see also Hogan and McCrea (1974), McCrea and O'Brien (1978), and O'Brien (1979)].

The discovery of the binary neutron star system PSR 1913+16 in 1974 provided the impetus for renewed interest in the post-Newtonian approximation schemes, since it was the first isolated gravitating system found in which general relativity plays a fundamental role in its evolution (Hulse and Taylor, 1975). In particular the indirect discovery of gravitational waves by observation of period shortening led to many fruitful studies of the equations of motion with gravitational radiation emission [see Damour (1983, 1987*a*, 1987*b*) for reviews]. The effect of gravitational radiation reaction appears in a very high order of approximation, relative to the Newtonian approximation, in the equations of motion of isolated bodies. Ehlers *et al.* (1976) [see also Pati and Will (2000)] critically discussed the foundation of the so-called quadrupole formula for the radiation reaction. There have been many studies demonstrating the validity of the quadrupole formula: Damour (1983), Anderson and DeCanio (1975), McCrea (1981), Futamase (1983), Isaacson *et al.* (1984), Kates (1980*a*, 1980*b*), Kerlick (1980*a*, 1980*b*), Schäfer (1985, 1986), Schutz (1980, 1984), and Walker and Will (1980*a*, 1980*b*). Notably, Damour (1982) established the formula for compact binaries with the help of the "dominant Schwarzschild condition"(Damour, 1983) while Blanchet and Damour (1989) established it for general fluid systems.

With the advent of kilometer-size interferometric gravitational wave detectors such as LIGO (Abramovici *et al.*, 1992), VIRGO (Bradaschia *et al.*, 1990), GEO (Danzmann, 1992) and TAMA (Kuroda, 1997) a further challenge and stimulus to equations of motion research arrived. This is because the most promising candidates for sources of gravitational waves are coalescing binary neutron stars. The significant features of this phenomenon are its frequency, expected once every year to once every few hundred years (Burgay *et al.*, 2003; Kalogera *et al.*, 2004), and the wave profile or waveform associated with it which can be predicted with greater accuracy than that of any other sources of gravitational waves (Abramovici *et al.*, 1992; Thorne, 1994; Will, 1994). Information carried by gravitational waves provides knowledge of various physical parameters associated with neutron stars (Cutler and Thorne, 2002) and also of cosmological parameters [see Finn (1996), Markovic (1993), Schutz (1986, 2002) and Wang *et al.* (1996)], provided that a sufficiently accurate comparison can be made between the observed and theoretical waveforms during the inspiraling phase of the binary star system when the orbital separation is much larger than the radii of the component stars. This is the epoch in the history of the binary star system when the post-Newtonian approximation can be applied to evaluate theoretical profiles for gravitational waves. The calculations however have to be carried out to a very high degree of accuracy relative to the Newtonian approximation in order to arrive at useful results and this has stimulated some ingenious ways of carrying them out.

1.2 Equations of motion of small black holes

Another aspect of the problem of motion of extended bodies in general relativity is to consider the equations of motion of small black holes. These are extended bodies too and on account of their existence we may say that in general relativity there is no such thing as a point source of a gravitational field. An isolated source may look point-like at large distance, and it may be useful to treat it as a point, but that is a separate issue. For this reason we should be surprised to find in general relativity infinities of the type encountered in classical electrodynamics. There a point electron has infinite self-energy, and in addition can "run away" or accelerate to the speed of light under the action of no external field and can even exhibit "pre-acceleration" or violation of causality when subjected to an impulsive electric field, for example (Dirac, 1938). By contrast a small Reissner–Nordström black hole, or a small Schwarzschild black hole, which are black holes experiencing no external fields, neither contain infinities nor do they in any sense exhibit any of the peculiar motions mentioned above. This raises the question as to what is the explanation for these contrasting situations. For the case of a charged black hole we take the view that the natural way to combine charge e and mass m is to use the Einstein–Maxwell field theory. This may seem an obvious statement. However in the original important work of Dirac (1938) the only field theory used was Maxwell's electromagnetic theory in Minkowskian space-time to describe the electromagnetic field of the particle having charge e. The mass m of the point charge was then brought into play with the use of a 4-momentum balancing argument, relating the field 4-momentum (involving e) crossing a small world tube, surrounding a segment of the world-line of the charge, to the change in particle 4-momentum (involving m) between the ends of the world tube. The extension of Dirac's argument to curved space-time carried out by DeWitt and Brehme (1960) utilizes the same approach with the field theory now in use being Maxwell's theory on curved space-time. Incidentally insisting on the use of the Einstein–Maxwell theory eliminates the concept of an impulsive electric field since the electromagnetic energy tensor associated with such a field is mathematically meaningless.

Having decided to make use of the Einstein–Maxwell field theory to study the motion of a small charged black hole moving in external electromagnetic and gravitational fields, or of an uncharged black hole moving in an external gravitational field, we focus on the behavior of the wave fronts of the radiation, electromagnetic and/or gravitational, produced by the motion to obtain information about this motion. The key issues are that (a) the wave fronts should have no singular points and (b) they should be smoothly deformed spheres near the small black hole. A singular point on a wave front translates in space-time to a singular null geodesic in the null hypersurface history of the wave front. Such "directional" singularities contradict the notion of an isolated body in general relativity. In a technical sense, explained in Chapter 6, the electromagnetic and/or gravitational fields near the small black hole resemble the Coulomb and/or Reissner–Nordström fields, which are spherically symmetric, and thus radiation emitted on account of the motion would naturally be expected to have spherical wave fronts near the black hole and suffer distortion due to the external fields as the radiation propagates. These basic ideas are best enunciated, albeit in the context of space-times which are more restricted than those required here, by Robinson and Robinson (1972) in a seminal and challenging paper with the disconcertingly modest title of "Equations of motion in the linear approximation".

2 Foundations of the post-Newtonian approximation

The strong field point particle limit is defined, the surface integral approach to the evaluation of the equations of motion is described and the scalings of the matter and field variables on an initial space-like hypersurface are discussed. The Newtonian velocity–momentum relation and the Newtonian equations of motion for extended bodies are re-visited and the equations of motion in the first post-Newtonian (PN) approximation are obtained and their derivation carefully analyzed. The spin–orbit coupling force, the quadrupole–orbit coupling force, the spin–spin coupling force and the spin geodesic precession equation are calculated to lowest order. These calculations are described in Appendix B which, in addition to Appendix A, is directly relevant to this chapter.

2.1 Setting the scene

It is well known that the post-Newtonian approximation to general relativity is important in analyzing a number of relativistic problems, such as the equations of motion of binary pulsars (Blandford and Teukolsky, 1976*a*; Damour and Taylor, 1991; Epstein, 1977*a*; Hulse and Taylor, 1975), solar-system tests of general relativity [see Will (1987, 1993, 1994, 2006)] and gravitational radiation reaction (Burke, 1971; Chandrasekhar and Esposito, 1970). Any approximation scheme requires one or more small parameters characterizing the physical system under consideration. A typical parameter which most approximation schemes adopt is the magnitude of the metric deviation from a background metric. In particular if the background is Minkowskian space-time, and there is no other parameter, the scheme is sometimes called the post-Minkowskian approximation in the sense that the constructed space-time reduces to Minkowskian space-time in the limit in which the parameter tends to zero. This limit is called the weak field limit. In the case of the post-Newtonian approximation the background space-time is also Minkowskian space-time, but there is another small parameter taken to be the typical velocity of the system divided by the speed of light. We introduce a dimensionless parameter ϵ to express the *slowness* of the system. These two parameters (the deviation from the flat metric and the velocity) are required to have a certain relation in the following sense. As the gravitational field gets weaker, all velocities and forces characteristic of the material systems become smaller, in order to permit the weakening of gravity to remain an important effect in the system's dynamics. For example in the case of a binary system the typical velocity would be the orbital velocity $v/c \sim \epsilon$ (where c is the speed of light in a vacuum) and the deviation of the space-time metric from the flat metric would be expressed via the Newtonian potential Φ (say). These variables are then related by $\Phi/c^2 \sim v^2/c^2 \sim \epsilon^2$, which guarantees that the system is bounded by its own gravity.

In the post-Newtonian approximation, the equations of general relativity take the form of Newton's equations in an appropriate limit as $\epsilon \to 0$. Such a limit is called the Newtonian limit and it will be the basis of constructing the post-Newtonian approximation. However, the limit is not in any sense trivial since it may be thought of as two limits tied together as just described. It is also worth noting that the Newtonian limit cannot be uniform everywhere for all time. For example any compact binary system, no matter how weak the gravity between its components and how slow its orbital motion, will eventually spiral together due to backreaction from the emission of gravitational waves. As a result the effects of its Newtonian gravity will be swamped by those of its gravitational waves. This will mean that higher-order effects of the post-Newtonian approximation eventually dominate the lowest-order Newtonian dynamics and thus if the post-Newtonian approximation is not carefully constructed this effect can lead to many formal problems such as divergent integrals (Ehlers *et al.*, 1976). It has been shown that such divergences can be avoided by carefully defining the Newtonian limit (Futamase, 1983). Moreover the use of such a limit provides us with a strong indication that the post-Newtonian hierarchy is an asymptotic approximation to general relativity (Futamase and Schutz, 1983). Therefore we shall first discuss the Newtonian limit and how to construct the post-Newtonian hierarchy. We follow the formulation developed by Futamase and Schutz (1983). Alternative formulations can be found in Daucourt (1997), Ehlers (1997), and Rendall (1992).

2.2 Newtonian limit and asymptotic Newtonian sequence

This formulation is based on the observation that any asymptotic approximation of any theory needs a sequence of solutions of the basic equations of the theory (Stewart and Walker, 1974; Wald, 1984). Thus if we write the equations in abstract form as

$$E(g) = 0, \tag{2.1}$$

for an unknown function g, we would like to have a one-parameter (or possibly multi-parameter) family of solutions,

$$E(g(\lambda)) = 0, \tag{2.2}$$

where λ is some parameter. Asymptotic approximation then means that a function $f(\lambda)$ approximates $g(\lambda)$ to order λ^p if $|f(\lambda) - g(\lambda)|/\lambda^p \to 0$ as $\lambda \to 0$. We choose the sequence of solutions with appropriate properties in such a way that the properties reflect the character of the system under consideration.

We shall formulate the post-Newtonian approximation according to this general notion. We aim to have an approximation that applies to the systems whose motions are described almost by Newtonian theory. Thus we need a sequence of solutions of the Einstein equations parameterized by ϵ (the typical velocity of the system divided by the speed of light) which has Newtonian character as $\epsilon \to 0$.

The Newtonian character is most conveniently described by the following scaling law. The Newtonian equations involve six variables, namely the density ρ, the pressure P, the gravitational potential Φ, and the velocity v^i, $(i = 1, 2, 3)$:

$$\nabla^2 \Phi - 4\pi\rho = 0, \tag{2.3}$$

$$\partial_t \rho + \nabla_i(\rho v^i) = 0, \tag{2.4}$$

$$\rho\partial_t v^i + \rho v^j \nabla_j v^i + \nabla^i P + \rho\nabla^i \Phi = 0, \tag{2.5}$$

supplemented by an equation of state. For simplicity we have considered a perfect fluid. It can be seen that the variables $\rho(x^i, t), P(x^i, t), \Phi(x^i, t)$ and $v^i(x^j, t)$ obeying the above equations satisfy the following scaling law:

$$\begin{aligned} \rho(x^i, t) &\to \epsilon^2 \rho(x^i, \epsilon t), \\ P(x^i, t) &\to \epsilon^4 P(x^i, \epsilon t), \\ v^i(x^k, t) &\to \epsilon v^i(x^k, \epsilon t), \\ \Phi(x^i, t) &\to \epsilon^2 \Phi(x^i, \epsilon t). \end{aligned} \tag{2.6}$$

One can easily understand the meaning of this scaling since the parameter ϵ is the *slowness* introduced earlier. The magnitude of the gravitational potential will be of order ϵ^2 because of the balance between gravity and the centrifugal force. The scaling of the time variable expresses the fact that the weaker gravity is $(\epsilon \to 0)$ the longer is the time scale. Thus we wish to have a sequence of solutions of Einstein's equations which has the above scaling as $\epsilon \to 0$. We shall also take the point of view that the sequence of solutions is determined by the appropriate sequence of *initial data*. This latter viewpoint has a practical advantage because there will be no *solutions* of Einstein's equations which satisfy the above scaling (2.6) even as $\epsilon \to 0$. This is because Einstein's equations are nonlinear in the field variables, so it will not be possible to enforce the scaling everywhere in space-time. We shall therefore impose it only on the initial data for the solution of the sequence. Following a general discussion on the formulation of the post-Newtonian approximation, independent of any initial value formalism, we will then present the concrete treatment in harmonic coordinates. Harmonic coordinates are used because of their popularity and the fact that they have some advantages in the generalization to systems with strong internal gravity.

As initial data for the matter we take the same data set as in the Newtonian case, namely the density ρ, the pressure P, and the coordinate velocity v^i. In applications we usually assume a simple equation of state which relates the pressure to the density. The initial data for the gravitational field are $g_{\mu\nu}, \partial g_{\mu\nu}/\partial t$. Since the field equations of general relativity constitute an overdetermined system of partial differential equations there will be constraint equations among the initial data for the field. We shall write the free data for the field as (Q_{ij}, P_{ij}) whose explicit forms depend on the coordinate condition one assumes. In any coordinates we shall assume these data for the field vanish since we are interested in the evolution of an isolated system by its own gravitational interaction. It is expected that this choice corresponds to the absence of radiation far from the source. Thus we choose the following initial data dictated by the Newtonian scaling:

$$\begin{aligned}\rho(t=0,x^i,\epsilon) &= \epsilon^2 a(x^i),\\ P(t=0,x^i,\epsilon) &= \epsilon^4 b(x^i),\\ v^i(t=0,x^k,\epsilon) &= \epsilon c^i(x^k),\\ Q_{ij}(t=0,x^i,\epsilon) &= 0,\\ P_{ij}(t=0,x^i,\epsilon) &= 0,\end{aligned} \tag{2.7}$$

where the functions a, b, and c^i are C^∞ functions with compact support contained entirely within a sphere of a finite radius.

Corresponding to the above data, we have a one-parameter family of space-times parameterized by ϵ. It may be helpful to visualize the family as a fiber bundle, with base space the real line R (coordinate ϵ) and fiber the space-time R^4 (coordinates t, x^i). The fiber $\epsilon = 0$ is Minkowskian space-time since it is defined by zero data. In the following we shall assume that the solutions are sufficiently smooth functions of ϵ for small $\epsilon \neq 0$. We wish to take the limit $\epsilon \to 0$ along the sequence. The limit is, however, not unique and is defined by giving a smooth nowhere vanishing vector field on the fiber bundle which is nowhere tangent to each fiber (Stewart and Walker, 1974; Geroch, 1969). The integral curves of the vector field give a correspondence between points in different fibers, namely events in different space-times with different values of ϵ. Bearing in mind the Newtonian scaling of the time variable in the limit, we introduce the Newtonian dynamical time

$$\tau = \epsilon t, \tag{2.8}$$

and define the integral curve as the curve on which τ and x^i remain constant (in the sequel the coordinates (τ, x^i) will be referred to as the near zone coordinates). In fact if we take the limit $\epsilon \to 0$ along this curve, the orbital period of the binary system with $\epsilon = 0.01$ is 10 times that of the system with $\epsilon = 0.1$ as expected from the Newtonian scaling. This is what we define as the Newtonian limit. Notice that this map never reaches the fiber $\epsilon = 0$ (Minkowski space-time). There is no pure vacuum Newtonian limit as expected.

From now on we assume the existence of a sequence of solutions constructed from the initial data and satisfying the above scaling with respect to ϵ. We shall call such a sequence a regular asymptotically Newtonian sequence. We have to make further mathematical assumptions about the sequence in order to carry out explicit calculations. We note that to formally justify our assumptions would require a deep understanding of the existence and uniqueness properties of the Cauchy problem for Einstein's equations with perfect fluids of compact support which is not available at present.

2.3 Post-Newtonian hierarchy

We now define the Newtonian, post-Newtonian (PN), and higher approximations of various quantities as the appropriate higher tangents of the corresponding quantities to the above integral curve at $\epsilon = 0$. For example the hierarchy of approximations for the space-time metrics can be expressed as follows:

$$g_{\mu\nu}(\epsilon, \tau, x^i) = g_{\mu\nu}(0, \tau, x^i) + \epsilon(\mathcal{L}_V g_{\mu\nu})(0, \tau, x^i) \\ + \frac{1}{2}\epsilon^2 \left(\mathcal{L}_V^2 g_{\mu\nu}\right)(0, \tau, x^i) + \cdots + \frac{\epsilon^n}{n!}\left(\mathcal{L}_V^n g_{\mu\nu}\right)(0, \tau, x^i) \\ + R_{n+1}, \tag{2.9}$$

where $\mathcal{L}_V$ is the Lie derivative with respect to the tangent vector to the curves defined above and the remainder term $R^{\mu\nu}_{n+1}$ is

$$R^{\mu\nu}_{n+1} = \frac{\epsilon^{n+1}}{(n+1)!}\int_0^1 d\ell\,(1-\ell)^{n+1}\left(\mathcal{L}_V^{n+1} g_{\mu\nu}\right)(\ell\epsilon, \tau, x^i). \tag{2.10}$$

Taylor's theorem guarantees that the series is an asymptotic expansion about $\epsilon = 0$ under certain assumptions mentioned above. It may be useful to point out that the above definition of the approximation scheme may be formulated purely geometrically in terms of a jet bundle.

This post-Newtonian hierarchy gives us an asymptotic series in which each term in the series is manifestly finite. This is based on the ϵ dependence of the domain of dependence of the field point (τ, x^k). The region is finite with finite values of ϵ, and the diameter of the region increases like ϵ^{-1} as $\epsilon \to 0$. Without this linkage of the region with the expansion parameter ϵ the PN approximation leads to divergences in the higher orders. This is closely related to the retarded expansion where it is assumed that the slow motion assumption enables one to Taylor expand the retarded integrals in retarded time such as

$$\int dr\, f(\tau - \epsilon r, \dots) = \int dr\, f(\tau, \dots) - \epsilon \int dr\, r f(\tau, \dots)_{,\tau} + \dots, \tag{2.11}$$

and to assign the second term here to a higher order because of the explicit factor of ϵ. This is incorrect because $r \to \epsilon^{-1}$ as $\epsilon \to 0$ and thus ϵr is not uniformly small in the Newtonian limit. Only if the integrand falls off sufficiently fast can the retardation be ignored. This happens in the lower-order PN terms. But at some higher order there appear many terms which do not fall off sufficiently fast because of the nonlinearity of Einstein's equations. This is the reason why the formal post-Newtonian approximation produces divergent integrals. It turns out that such a divergence appears at 4PN order, indicating a breakdown of the PN approximation in harmonic coordinates (Blanchet and Damour, 1988). Historically it was said that in harmonic coordinates there is a "breakdown" at 3PN order [see for example Kerlick (1980*a*)] however the divergences at 3PN order may be gauge effects. This sort of divergence can be eliminated if we remember that the upper bound of the integral depends on ϵ as ϵ^{-1}. Thus we would get something like $\epsilon^n \ln \epsilon$ instead of $\epsilon^n \ln \infty$ in the usual approach. This shows that the asymptotic Newtonian sequence is not differentiable in ϵ at $\epsilon = 0$, but there is no divergence in the expansion and it has still an asymptotic approximation in ϵ that involves logarithms.

In addition to the initial value formulation method (Futamase, 1983; Futamase and Schutz, 1983) mentioned above, various other methods have been proposed to solve the problem of the divergent integrals. It is known that a higher-order post-Newtonian metric does not respect the asymptotically flat condition. This does not mean that the post-Newtonian approximation is useless at such a high order. The problem is related to the fact that a simple post-Newtonian iteration is meaningful only in the near zone – about one wavelength distance away from the material source – and is not useful outside of the near zone, called the far zone, where the wave effect (retardation effect) is manifest. So roughly speaking if a far-zone metric satisfying proper boundary conditions at infinity is solved so that we have a boundary condition to the field equations for a post-Newtonian metric in the buffer zone, we can find a post-Newtonian metric which is meaningful in the sense that it respects the correct behavior at the near-zone boundary.

Blanchet and Damour have developed a systematic approach of a matched asymptotic expansion. They solved the far-zone metric using a multipolar post-Minkowskian expansion. The far-zone metric satisfies a stationarity condition and is parameterized by radiative multipole moments. On the other hand they solve a post-Newtonian near-zone metric up to a homogeneous solution. They then establish an association between those radiative multipole moments and the source multipole moments that characterize a post-Newtonian near-zone metric to fix the homogeneous solution, and find a post-Newtonian metric which satisfies the correct behavior in the buffer zone.

Will and Wiseman have developed the DIRE method (Pati and Will, 2000; Will, 1999; Will and Wiseman, 1996) where they split the integration region in the retarded integral into two regions – the near zone and the far zone. The near-zone metric is solved by a post-Newtonian expansion. The retarded integral over the far zone is directly evaluated with the assumption of sufficient stationarity of the system in the infinite past. Both the Blanchet–Damour method and the Will–Wiseman method have been proved to give a physically equivalent result (Blanchet, 2006). For the computation here of the 3PN equations of motion we will use the Will and Wiseman method to solve the problem of the breakdown of the post-Newtonian approximation in the near zone (Futamase and Itoh, 2007).

2.4 A calculation using harmonic coordinates

To illustrate the formalism introduced above we give an explicit calculation in harmonic coordinates. The reduced Einstein equations in harmonic coordinates take the form

$$\tilde{g}^{\alpha\beta}\tilde{g}^{\mu\nu}{}_{,\alpha\beta} = 16\pi\Theta^{\mu\nu} - \tilde{g}^{\mu\alpha}{}_{,\beta}\tilde{g}^{\nu\beta}{}_{,\alpha}, \tag{2.12}$$

$$\partial_\mu[\tilde{g}^{\mu\nu}\partial_\nu x^\alpha] = 0, \tag{2.13}$$

where

$$\tilde{g}^{\mu\nu} = (-g)^{1/2}g^{\mu\nu}, \tag{2.14}$$

$$\Theta^{\alpha\beta} = (-g)\left(T^{\alpha\beta} + t^{\alpha\beta}_{\rm LL}\right), \tag{2.15}$$

and $t_{\rm LL}^{\mu\nu}$ is the Landau–Lifshitz pseudotensor (Landau and Lifshitz, 1975). In this section we shall choose an isentropic perfect fluid for $T^{\alpha\beta}$, which is enough for most applications, and thus

$$T^{\alpha\beta} = (\rho + \rho\Pi + P)u^{\alpha}u^{\beta} + Pg^{\alpha\beta}, \tag{2.16}$$

where ρ is the rest mass density, Π the internal energy, P the pressure, and u^{μ} the 4-velocity of the fluid with normalization

$$g_{\alpha\beta}u^{\alpha}u^{\beta} = -1. \tag{2.17}$$

The conservation of energy and momentum is expressed as

$$\nabla_{\beta}T^{\alpha\beta} = 0. \tag{2.18}$$

Defining the gravitational field variable as

$$h^{\mu\nu} = \eta^{\mu\nu} - (-g)^{1/2}g^{\mu\nu}, \tag{2.19}$$

where $\eta^{\mu\nu}$ is the Minkowski metric, the reduced Einstein equations (2.12) and the gauge condition (2.13) take the following form:

$$(\eta^{\alpha\beta} - h^{\alpha\beta})h^{\mu\nu}{}_{,\alpha\beta} = -16\pi\Theta^{\mu\nu} + h^{\mu\alpha}{}_{,\beta}h^{\nu\beta}{}_{,\alpha}, \tag{2.20}$$

$$h^{\mu\nu}{}_{,\nu} = 0. \tag{2.21}$$

Thus the characteristics are determined by the operator $(\eta^{\alpha\beta} - h^{\alpha\beta})\partial_{\alpha}\partial_{\beta}$, and so the light cone deviates from that in the flat space-time. We may use this form of the reduced Einstein equations in the calculation of the waveform far from the source because this deviation plays a fundamental role there (Asada and Futamase, 1997*a*). However, in studying the gravitational field near the source it is not necessary to consider the deviation of the light cone from the flat case and thus it is convenient to use the following form of the reduced Einstein equations (Anderson and DeCanio, 1975):

$$\eta^{\mu\nu}h^{\alpha\beta}{}_{,\mu\nu} = -16\pi\Lambda^{\alpha\beta}, \tag{2.22}$$

where

$$\Lambda^{\alpha\beta} = \Theta^{\alpha\beta} + \chi^{\alpha\beta\mu\nu}{}_{,\mu\nu}, \tag{2.23}$$

$$\chi^{\alpha\beta\mu\nu} = (16\pi)^{-1}(h^{\alpha\nu}h^{\beta\mu} - h^{\alpha\beta}h^{\mu\nu}). \tag{2.24}$$

Equations (2.21) and (2.22) together imply the conservation law

$$\Lambda^{\alpha\beta}{}_{,\beta} = 0. \tag{2.25}$$

We shall take as our variables the set $\{\rho, P, v^{i}, h^{\alpha\beta}\}$ with the definition

$$v^{i} = u^{i}/u^{0}. \tag{2.26}$$

The time component of the 4-velocity u^{0} is determined from (2.17). To make a well-defined system of equations we must add the conservation law for the number density n, which is some function of the density ρ and pressure P:

$$\nabla_{\alpha}(nu^{\alpha}) = 0. \tag{2.27}$$

Equations (2.25) and (2.27) imply that the flow is adiabatic. The role of the equation of state is played by the arbitrary function $n(\rho, p)$.

Initial data for the above set of equations consists of $h^{\alpha\beta}, h^{\alpha\beta}{}_{,0}, \rho, P$, and v^i, but not all these data are independent because of the existence of the constraint equations. Equations (2.21) and (2.22) imply the four constraint equations among the initial data for the field,

$$\Delta h^{\alpha 0} + 16\pi\Lambda^{\alpha 0} - \delta^{ij} h^{\alpha}_{i,j}{}^{,0} = 0, \tag{2.28}$$

where Δ is the Laplacian in the flat space. We shall choose h^{ij} and $h^{ij}{}_{,0}$ as free data and solve (2.28) for $h^{\alpha 0}(\alpha = 0, \ldots, 3)$ and (2.21) for $h^{\alpha 0}{}_{,0}$. Of course these constraints cannot be solved explicitly, since $\Lambda^{\alpha 0}$ contains $h^{\alpha 0}$, but they can be solved iteratively as explained below. As discussed above, we shall assume that the free data h^{ij} and $h^{ij}{}_{,0}$ for the field vanish. One can show that such initial data satisfy the O'Murchadha and York criterion for the absence of radiation far from the source (O'Murchadha and York, 1974).

In carrying out the calculation we find that it is convenient to use an expression with explicit dependence on ϵ. The harmonic condition allows us to have such an expression in terms of the retarded integral

$$h^{\mu\nu}(\epsilon, \tau, x^i) = 4 \int_{C(\epsilon,\tau,x^i)} d^3y\, \Lambda^{\mu\nu}(\tau - \epsilon r, y^i, \epsilon)/r + h^{\mu\nu}_{\mathrm{H}}(\epsilon, \tau, x^i), \tag{2.29}$$

where $r = |y^i - x^i|$ and $C(\epsilon, \tau, x^i)$ is the past flat light cone of the event (τ, x^i) in the space-time given by ϵ, truncated where it intersects with the initial hypersurface $\tau = 0$. $h^{\mu\nu}_{\mathrm{H}}$ is the unique solution of the homogeneous wave equation in flat space-time,

$$\eta^{\alpha\beta} h^{\mu\nu}_{H\,,\alpha\beta} = 0. \tag{2.30}$$

$h^{\mu\nu}_{\mathrm{H}}$ evolves from a given initial data on the $\tau = 0$ initial hypersurface which are subject to the constraint equations (2.28). The explicit form of $h^{\mu\nu}_{\mathrm{H}}$ is available via the Poisson formula [see for example Sobolev (1989)]

$$h^{\mu\nu}_{\mathrm{H}}(\tau, x^i) = \frac{\tau}{4\pi} \oint_{\partial C(\tau,x^i)} h^{\mu\nu}{}_{,\tau}(0, y^i)\, d\Omega_y + \frac{1}{4\pi}\frac{\partial}{\partial\tau}\left[\tau \oint_{\partial C(\tau,x^i)} h^{\mu\nu}(0, y^i)\, d\Omega_y\right]. \tag{2.31}$$

The zero in the argument of each integrand indicates that the integrands are to be evaluated at $\tau = 0$. We shall henceforth ignore the homogeneous solutions because they play no role. Because of the ϵ dependence of the integration region the domain of integration is finite as long as $\epsilon \neq 0$ and the diameter increases like ϵ^{-1} as $\epsilon \to 0$.

Given the formal expression (2.29) in terms of initial data (2.7), we can take the Lie derivatives and evaluate these derivatives at $\epsilon = 0$. The Lie derivative is a partial derivative with respect to ϵ in the coordinate system for the fiber bundle given by (ϵ, τ, x^i). Accordingly one should convert all the time indices to τ indices. For example $T^{\tau\tau} = \epsilon^2 T^{tt}$, which is of order ϵ^4 since $T^{tt} \sim \rho$ is of order ϵ^2. Similarly the other components of the stress–energy tensor $T^{\tau i} = \epsilon T^{ti}$ and T^{ij} are of order ϵ^4. Thus we expect that the first of the non-vanishing Lie derivatives of (2.29) will be the fourth derivative. In fact we find

$$ {}_4h^{\tau\tau}(\tau, x^i) = 4\int_{R^3} \frac{{}_2\rho(\tau, y^i)}{r}\, d^3y, \tag{2.32} $$

$$ {}_4h^{\tau i}(\tau, x^i) = 4\int_{R^3} \frac{{}_2\rho(\tau, y^k){}_1v^i(\tau, x^k)}{r}\, d^3y, \tag{2.33} $$

$$ {}_4h^{ij}(\tau, x^k) = 4\int_{R^3} \frac{{}_2\rho(\tau, y^k){}_1v^i(\tau, y^k){}_1v^j(\tau, y^k) + {}_4t^{ij}_{\mathrm{LL}}(\tau, y^k)}{r}\, d^3y, \tag{2.34} $$

$$ \tag{2.35} $$

where we have adopted the notation

$$ {}_nf(\tau, x^i) = \frac{1}{n!}\lim_{\epsilon\to 0}\frac{\partial^n}{\partial\epsilon^n} f(\epsilon, \tau, x^i), \tag{2.36} $$

and

$$ {}_4t^{ij}_{\mathrm{LL}} = \frac{1}{64\pi}\left({}_4h^{\tau\tau,i}{}_4h^{\tau\tau,j} - \frac{1}{2}\delta^{ij}{}_4h^{\tau\tau,k}{}_4h^{\tau\tau}{}_{,k}\right). \tag{2.37} $$

In the above calculation we have taken the point of view that $h^{\mu\nu}$ is a tensor field, defined by giving its components in the assumed harmonic coordinates as the difference between the tensor density $\sqrt{-g}g^{\mu\nu}$ and $\eta^{\mu\nu}$. The conservation law (2.25) also has its first nonvanishing derivatives at this order, which are

$$ {}_2\rho_{,\tau} + ({}_2\rho_1v^i)_{,i} = 0, \tag{2.38} $$

$$ ({}_2\rho_1v^i)_{,\tau} + ({}_2\rho_1v^i{}_1v^j)_{,j} + {}_4P^{,i} - \frac{1}{4}{}_2\rho_4h^{\tau\tau,i} = 0. \tag{2.39} $$

Equations (2.32), (2.38), and (2.39) correspond to the equations of the Newtonian theory of gravity. Thus the lowest nonvanishing derivative with respect to ϵ is indeed Newtonian theory, and the 1PN and 2PN equations emerge from the sixth and eighth derivatives, respectively, in the conservation law (2.25). At the next derivative the quadrupole radiation reaction term emerges.

2.5 The strong field point particle limit

In a relativistic compact binary system in the inspiraling phase each star is well approximated by a point particle with a few low-order multipole moments. This is because, in the inspiraling phase, the stellar size divided by the orbital separation is much smaller than unity. If we wish to apply the post-Newtonian approximation to the inspiraling phase of binary neutron stars then the strong internal gravity must be taken into account. The usual post-Newtonian approximation explicitly assumes the weakness of the gravitational field everywhere, including inside the material source. It is argued by applying the strong equivalence principle that the external gravitational field, which governs the orbital motion of the binary system, is independent of the internal structure of the components up to tidal interaction. Thus it is expected that the results obtained under the assumption of weak gravity also apply for the case of a neutron star binary. However observational evidence for the strong equivalence principle exists only for systems with weak gravity [see Will (1987, 1993, 1994, 2006)], while no evidence is currently available for the case of strong internal gravity.

The theory of extended objects in general relativity (Dixon, 1979) is still at too early a stage of development for application to realistic systems. The matched asymptotic expansion technique has been used to treat a system with strong gravity in certain situations (Damour, 1983; D'Eath, 1975*a*, 1975*b*; Kates, 1980*a*, 1980*b*). Another way to handle strong internal gravity is by the use of Dirac delta functions in the description of the source with a fixed mass (Einstein *et al.*, 1938). When the nonlinearity of Einstein's equations is taken into consideration this approach becomes mathematically meaningless. Physically there is no such thing as a point source in general relativity because of the existence of black holes. Before a body shrinks to a point it forms a black hole whose size is fixed by its mass. This does not mean that a useful approximation to a point source cannot be constructed. This can be achieved by shrinking a body of mass M and radius R in such a way that the compactness M/R, or strength of the internal field, remains fixed. Such a limit can be fitted nicely into the concept of the regular asymptotic Newtonian sequence defined above because there the mass also scales along the sequence of solutions. In fact if we take the masses of two stars as M, and the separation between the two stars as L, then $\Phi \sim M/L$. Thus the mass M scales as ϵ^2 if we fix the separation. We are assuming that the density scales as ϵ^2 to guarantee this scaling for the mass while keeping the size of the body fixed. Now if we shrink the size as ϵ^2, to keep the compactness of each component, then the density should scale as ϵ^{-4}. We shall call such a scheme the strong field point particle limit since the limit preserves the strength of the internal gravity. This is sometimes referred to as an extended body approach since we set ϵ to some finite value when we apply our method to real binary systems. These considerations suggest the following initial data to define a regular asymptotic Newtonian sequence which describes a nearly Newtonian system with strong internal gravity (Futamase, 1987): The initial data are two uniformly rotating fluids with compact spatial support whose stress–energy tensor and size scales as ϵ^{-4} and ϵ^2 respectively. We also assume that each of these fluid configurations would be a stationary equilibrium solution of Einstein's equations if the other were absent. This is necessary for the suppression of irrelevant internal motions of each star. Any remaining motions are tidal effects caused by the other body, which will be of order ϵ^6 smaller than the internal self-force. These data allow us to use the Newtonian time $\tau = \epsilon t$ as a natural time coordinate everywhere including the interior region of the stars.

When we consider multipole moments the scaling $R = O(\epsilon^2)$ enables us to incorporate the multipole expansion of the stars into the post-Newtonian approximation. For inspiraling compact binaries the tidally and rotationally induced quadrupole moments are too small to affect the binary orbital motion. The time to coalesce is too short for the binary to be tidally locked and co-rotate (Bildsten and Cutler, 1992; Cutler *et al.*, 1993; Kochanek, 1992). The phase shift due to the quadrupole–orbit couplings may be negligible in the LIGO bandwidth (Cutler *et al.*, 1993). However to detect gravitational waves from inspiraling binaries highly accurate prior knowledge of the binary orbital motion is required (typically this means a knowledge of the equations of motion at the 4PN order of approximation). The effect of quadrupole moments on the orbital motion can be of about the same order as that of spin–spin interactions (Poisson, 1998) while the spin–spin interactions appear at about 2.5PN order for slowly rotating stars. In addition, at the late inspiraling phase, the effect of the extendedness of the stars on the motion will be significant. Thus it is important to take the multipole moments of the stars into account in a way that is suitable for compact stars when deriving the equations of motion for an inspiraling compact binary.

2.6 Surface integrals and body zone

One way to evaluate the gravitational force acting on a star is to use the so-called volume integral approach. In Newtonian theory the gravitational force F_1^i on star 1 is given by

$$F_1^i = \int_{B_1} d^3x \, \rho \frac{\partial \Phi}{\partial x^i}, \tag{2.40}$$

where, as always, ρ is the mass density and Φ is the Newtonian potential. The integration volume B_1 includes star 1 but does not include star 2 (the companion star). To evaluate the above integral we must have some knowledge of the internal structure of the star. However, by means of the surface integral approach we can put off dealing with the internal structure until tidal effects affect the orbital motion. In Newtonian theory, using Poisson's equation and Gauss' law, we can rewrite the volume integral as the surface integral

$$F_1^i = -\oint_{\partial B_1} dS_j \, t^{ij}, \tag{2.41}$$

$$t^{ij} \equiv \frac{1}{4\pi}\left(\frac{\partial \Phi}{\partial x^i}\frac{\partial \Phi}{\partial x^j} - \frac{\delta^{ij}}{2}\frac{\partial \Phi}{\partial x^k}\frac{\partial \Phi}{\partial x^k}\right). \tag{2.42}$$

Note that the sphere $\partial \mathrm{B}_1$ has no intersection with star 1 or with star 2. Thus we can evaluate the gravitational force acting on star 1 without a knowledge of the internal structure of star 1. The surface integral approach was taken in general relativity by Einstein, Infeld, and Hoffmann (1938). They used Einstein's vacuum field equations and their method can be applied to any object including a black hole. We will use the surface integral approach but we will consider only regular objects such as neutron stars.

We must introduce the *body zone* to provide the surfaces for the surface integral approach. The scalings of R and M motivate us to define the body zone of star A ($A = 1, 2$) as $B_A \equiv \{x^i : |\vec{x} - \vec{z}_A(\tau)| < \epsilon R_A\}$ and the body zone coordinates of star A as $\alpha_A^i \equiv \epsilon^{-2}\left(x^i - z_A^i(\tau)\right)$. Here $z_A^i(\tau)$ is a representative point of star A such as its center of mass. The *body zone radius* R_A is an arbitrary length scale (much smaller than the orbital separation and not identical to the radius of the star) and is a constant ($dR_A/d\tau = 0$). In the body zone coordinates, the star does not shrink when $\epsilon \to 0$, while the boundary of the body zone goes to infinity.

It is appropriate to define the star's characteristic quantities such as its mass, spin etc., in terms of the body zone coordinates. Also the body zone provides us with a surface ∂B_A through which gravitational energy/momentum flows, which in turn results in the gravitational force acting on star A. Since the body zone boundary ∂B_A is far from the surface of star A we can evaluate the gravitational energy momentum flux over ∂B_A using the post-Newtonian gravitational field. In fact we shall express our equations of motion in terms of integrals over ∂B_A and so be able to evaluate them explicitly.

2.7 Scalings on the initial hypersurface

Following Futamase (1983), Futamase and Schutz (1983), and Schutz (1980) we use the initial value formulation to solve Einstein's equations. As initial data for the matter variables and the gravitational field we take a set of nearly stationary solutions of the exact Einstein equations representing two widely separated fluid balls, each of which rotates uniformly. We assume that these solutions are parameterized by ϵ and that the matter and field variables have the following scalings: the density scales as ϵ^{-4} (in the (t, x^i) coordinates), implied by the scalings of M and R. The scaling of the density suggests that the natural dynamical time (free fall time) η inside the star may be $\eta = \epsilon^{-2}t$. Thus if we cannot assume the almost stationary condition on the stars it is difficult to use the post-Newtonian approximation (Futamase, 1985*a*, 1987). We note that our formalism also applies to pulsating stars if the effect of pulsation is not important in the orbital motion. The velocity of stellar rotation is assumed to be $O(\epsilon)$. In other words we assume that the star rotates slowly and is pressure supported, meaning that the pressure P scales as ϵ^{-4}. If we consider polytropes, with $P(\rho) = K\rho^\gamma$ for instance, then the polytropic constant K must scale as $\epsilon^{4(\gamma-1)}$. No such sequence of solutions parametrized by ϵ seems to have been obtained so far. We simply assume the existence of such a sequence. With these assumptions the spin–orbit coupling force appears at 2PN order rather than the usual 1.5PN order. The slowly spinning motion assumption is not crucial. In fact it is straightforward to incorporate a rapidly spinning compact body into our formalism.

From these initial data we have the following scalings of the stress–energy tensor components $T_A^{\mu\nu}$ for star A in the body zone coordinates: $T_A^{\tau\tau} = O(\epsilon^{-2})$, $T_A^{\tau\underline{i}} = O(\epsilon^{-4})$, $T_A^{\underline{ij}} = O(\epsilon^{-8})$. The underlined indices mean that for any tensor A^i, $A^{\underline{i}} = \epsilon^{-2}A^i$. In (Itoh, Futamase, and Asada, 2000) the components of the stress–energy tensor of the matter in the near zone coordinates $T_N^{\mu\nu}$ have been transformed to $T_A^{\mu\nu}$ using the transformation from the near zone coordinates to the (generalized) Fermi normal coordinates at 1PN order (Ashby and Bertotti, 1986). It is difficult, however, to construct the (generalized) Fermi normal coordinates at a high post-Newtonian order and so we shall not make use of them in general. We simply assume that for $T_N^{\mu\nu}$ or rather $\Lambda_N^{\mu\nu}$, the source term of the relaxed Einstein equations (2.22),

$$T_N^{\tau\tau} = O(\epsilon^{-2}), \tag{2.43}$$

$$T_N^{\tau\underline{i}} = O(\epsilon^{-4}), \tag{2.44}$$

$$T_N^{\underline{ij}} = O(\epsilon^{-8}). \tag{2.45}$$

For the field variables on the initial hypersurface we assume that the field is of 2.5PN order except for the field determined by the constraint equations. Note that the radiation reaction effect on the stars first appears at 2.5PN order. Futamase showed that even if one takes the field of order 1PN, the initial value of the field does not affect the subsequent motion of the system up to 2.5PN order (Futamase, 1985*a*). Hence we expect that the initial value of the field does not affect the orbital motion of the system up to 3PN order, although a detailed calculation of this has not yet been carried out.

It is worth emphasizing that the initial value formulation has some advantages. Firstly by using it one can avoid the famous runaway solution difficulty in a radiation reaction problem. Secondly one can construct an initial condition on some space-like hypersurface rather than at past null infinity. To specify an initial condition for the field on past null infinity requires a prior knowledge about the space-time which is obtained through the time evolution of the field from the initial condition. The initial value formulation can give in a sense a realistic initial condition. In our universe there may be no past null infinity because of the big bang. An interesting initial condition is the statistical initial condition (Schutz, 1980). Here the binary system is in the background gravitational radiation bath for which only its statistical properties are known. For example the phase of the radiation is assumed to be random and irrelevant to the motion of the binary. The origins of the radiation are cosmological or related to the evolution of the system before the initial hypersurface. Then the expected time evolution of the binary system can be evaluated by letting the system evolve from a set of possible initial conditions and taking a statistical ensemble average over the initial conditions.

2.8 Newtonian equations for extended bodies

At this point in the argument it is useful to consider some equations in Newtonian gravitational theory relevant to extended bodies or stars. Such equations will provide helpful guidance for the development of the formalism. The basic equations we require are the equation of continuity, the Euler equation, and the Poisson equation, given respectively by

$$\frac{\partial \rho}{\partial \tau} + \frac{\partial \rho v^i}{\partial x^i} = 0, \tag{2.46}$$

$$\rho \frac{Dv^i}{D\tau} = \rho \frac{\partial \Phi}{\partial x_i}, \tag{2.47}$$

$$\Delta \Phi = 4\pi \rho. \tag{2.48}$$

We define the mass, the dipole moment, the quadrupole moment, and the momentum of the star A respectively by

$$m_A \equiv \int_{B_A} d^3x\,\rho, \tag{2.49}$$

$$D^i_A \equiv \int_{B_A} d^3x\,\rho\left(x^i - z^i_A\right), \tag{2.50}$$

$$I^{ij}_A \equiv \int_{B_A} d^3x\,\rho\left(x^i - z^i_A\right)\left(x^j - z^j_A\right), \tag{2.51}$$

$$P^i_A \equiv \int_{B_A} d^3x\,\rho v^i. \tag{2.52}$$

Here z^i_A is a representative point of the star A. The time derivative of the mass vanishes. Setting the time derivative of the dipole moment to zero gives the velocity–momentum relation and a definition of the center of mass. Thus

$$\frac{dD^i}{d\tau} = P^i_A - m_A v^i_A = 0, \tag{2.53}$$

where $v^i_A = dz^i_A/d\tau$. Using the velocity–momentum relation we calculate the time derivative of the momentum to arrive at

$$m_1\frac{d^2z^i_1}{d\tau^2} = F^i_1, \tag{2.54}$$

where F^i_1 is defined by Equation (2.41). The Newtonian potential can be expressed in terms of the mass and multipole moments as

$$\Phi = \sum_{A=1,2}\left[\frac{m_A}{|\vec{x}-\vec{z}_A|} + \frac{1}{2}I^{ij}_A\frac{\partial^2}{\partial x^i\,\partial x^j}\left(\frac{1}{|\vec{x}-\vec{z}_A|}\right)\right]. \tag{2.55}$$

Substituting Φ into (2.54) we obtain the equations of motion,

$$m_1\frac{d^2z^i_1}{d\tau^2} = \frac{\partial}{\partial z^i_1}\frac{m_1m_2}{|\vec{z}_1-\vec{z}_2|} + \frac{1}{2}\left(I^{ij}_1 + I^{ij}_2\right)\frac{\partial^3}{\partial z^i_1\partial z^j_2\partial z^k_2}\left(\frac{1}{|\vec{z}_1-\vec{z}_2|}\right). \tag{2.56}$$

Here we have ignored the mass multipole moments of the stars that are of higher order than the quadrupole moment. It is straightforward to formally include all the Newtonian mass multipole moments in the surface integral approach giving

$$m_1\frac{d^2z^i_1}{d\tau^2} = \sum_{p=0}\sum_{q=0}\frac{(-1)^{p+1}(2p+2q+1)!!}{p!q!}\frac{I_1^{\langle M_p\rangle}I_2^{\langle N_q\rangle}}{r_{12}^{p+q+2}}N^{\langle iM_pN_q\rangle}, \tag{2.57}$$

where $\vec{r}_{12} = \vec{z}_1 - \vec{z}_2$, $N^i = r^i_{12}/r_{12}$, $M_p = m_1m_2\ldots m_p$ is a collective index, $\langle\ldots\rangle$ denotes the symmetric-tracefree operation on the indices between the brackets, and $I_A^{M_p}$ are the Newtonian mass multipole moments of order $2p$.

2.9 Einstein's field equations

To extend the formalism, following Itoh (2000, 2001), to obtain the post-Newtonian equations of motion for compact binary systems we aim to express the equations of motion in terms of surface integrals over the body zone boundary where it is assumed that the metric slightly deviates from the flat metric $\eta^{\mu\nu} = \mathrm{diag}(-\epsilon^2, 1, 1, 1)$ (in the near zone coordinates (τ, x^i)). Thus we define a deviation field $h^{\mu\nu}$ as

$$h^{\mu\nu} \equiv \eta^{\mu\nu} - \sqrt{-g}\, g^{\mu\nu}, \tag{2.58}$$

where g is the determinant of the metric. This deviation field plays a central role in the remainder of this chapter and in Chapter 3. Henceforth we shall refer to the "field variables" $h^{\mu\nu}$ or, more simply, the "field" $h^{\mu\nu}$. Our $h^{\mu\nu}$ differs from the corresponding field in Blanchet *et al.* (1998) by a sign. Indices are raised or lowered by the flat auxiliary metric $\eta^{\mu\nu}$ unless otherwise stated. We impose the harmonic coordinate condition

$$h^{\mu\nu}{}_{,\nu} = 0, \tag{2.59}$$

where the comma denotes the partial derivative. In the harmonic gauge we can rewrite Einstein's equations in the relaxed form

$$\eta^{\mu\nu}\partial_\mu\partial_\nu = -16\pi\Lambda^{\mu\nu}, \tag{2.60}$$

where the left-hand side is the flat d'Alembertian and

$$\Lambda^{\mu\nu} \equiv \Theta^{\mu\nu} + \chi^{\mu\nu\alpha\beta}{}_{,\alpha\beta}, \tag{2.61}$$

$$\Theta^{\mu\nu} \equiv (-g)\left(T^{\mu\nu} + t^{\mu\nu}_{\mathrm{LL}}\right), \tag{2.62}$$

$$\chi^{\mu\nu\alpha\beta} \equiv \frac{1}{16\pi}(h^{\alpha\nu}h^{\beta\mu} - h^{\alpha\beta}h^{\mu\nu}). \tag{2.63}$$

Here $T^{\mu\nu}$ and $t^{\mu\nu}_{\mathrm{LL}}$ denote the stress–energy tensor of the stars and the Landau–Lifshitz pseudotensor respectively. The explicit form of $t^{\mu\nu}_{\mathrm{LL}}$ in the harmonic gauge is

$$\begin{aligned}(-16\pi g)t^{\mu\nu}_{\mathrm{LL}} &= g_{\alpha\beta}g^{\gamma\delta}h^{\mu\alpha}{}_{,\gamma}h^{\nu\beta}{}_{,\delta} + \frac{1}{2}g^{\mu\nu}g_{\alpha\beta}h^{\alpha\gamma}{}_{,\delta}h^{\beta\delta}{}_{,\gamma} - 2g_{\alpha\beta}g^{\gamma(\mu}h^{\nu)\alpha}{}_{,\delta}h^{\delta\beta}{}_{,\gamma} \\ &+\frac{1}{2}\left(g^{\mu\alpha}g^{\nu\beta} - \frac{1}{2}g^{\mu\nu}g^{\alpha\beta}\right)\left(g_{\gamma\delta}g_{\epsilon\zeta} - \frac{1}{2}g_{\gamma\epsilon}g_{\delta\zeta}\right)h^{\gamma\epsilon}{}_{,\alpha}h^{\delta\zeta}{}_{,\beta}.\end{aligned} \tag{2.64}$$

$\chi^{\mu\nu\alpha\beta}$ originates from our use of the flat d'Alembertian instead of the curved space d'Alembertian. Consistent with the harmonic condition, the conservation law is expressed as

$$\Lambda^{\mu\nu}{}_{,\nu} = 0. \tag{2.65}$$

Note that the divergence of $\chi^{\mu\nu\alpha\beta}$ vanishes identically due to the symmetry of its indices. Now we rewrite the relaxed Einstein equations in the integral form,

$$h^{\mu\nu}(\tau, x^i) = 4\int_{C(\tau,x^k)} d^3y\, \frac{\Lambda^{\mu\nu}(\tau - \epsilon|\vec{x} - \vec{y}|, y^k; \epsilon)}{|\vec{x} - \vec{y}|} + h^{\mu\nu}_{\mathrm{H}}(\tau, x^i), \tag{2.66}$$

where $C(\tau, x^k)$ means the past light cone emanating from the event (τ, x^k). $C(\tau, x^k)$ is truncated at the $\tau = 0$ hypersurface. $h^{\mu\nu}_{\rm H}$ is a homogeneous solution of the homogeneous wave equation in flat space-time (thus $\eta^{\alpha\beta}\partial_\alpha\partial_\beta\, h^{\mu\nu}_{\rm H} = 0$). $h^{\mu\nu}_{\rm H}$ evolves from given initial data on the $\tau = 0$ initial hypersurface. The explicit form of $h^{\mu\nu}_{\rm H}$ is available via the Poisson formula (Sobolev, 1989), namely,

$$h^{\mu\nu}_{\rm H}(\tau, x^i) = \frac{\tau}{4\pi}\oint_{\partial C(\tau,x^i)} h^{\mu\nu}{}_{,\tau}(0, y^i)\, d\Omega_y + \frac{1}{4\pi}\frac{\partial}{\partial\tau}\left[\tau\oint_{\partial C(\tau,x^i)} h^{\mu\nu}(0, y^i)\, d\Omega_y\right], \quad (2.67)$$

where, as before, the zero in the argument of each integrand indicates that the integrand is to be evaluated at $\tau = 0$.

To solve Einstein's equations we proceed as follows. First we split the integration region into two zones, the near zone and the far zone. The near zone is the region containing the gravitational wave source where the wave character of the gravitational radiation is not manifest. In other words in the near zone the retardation effects on the field are negligible. The near zone N incorporates the whole source system. The size of the near zone is just a little larger than one wavelength of the gravitational wave emitted by the source. For definiteness we take the near zone to be a sphere centered at some fixed point and enclosing the binary system. The radius of this sphere is set to be $\mathcal{R}/\epsilon$, where $\mathcal{R}$ is arbitrary but larger than the size of the binary and the wavelength of the gravitational radiation. The scaling of the near zone radius is derived from the ϵ dependence of the wavelength of the gravitational radiation emitted due to the orbital motion of the binary. Note that roughly speaking the frequency of such a wave is about twice the Keplerian frequency of the binary. The center of the near zone sphere could for example be taken to be the center of mass of the near zone. The region exterior to the near zone is the far zone where the retardation effects of the field are crucial.

At a near zone field point ${\rm P_N} \in N$ we write

$$h^{\mu\nu}({\rm P_N}) = h^{\mu\nu}_{P_N(N)} + h^{\mu\nu}_{P_N(F)} + h^{\mu\nu}_{\rm H}, \quad (2.68)$$

$$h^{\mu\nu}_{P_N(N)} \equiv 4\int_{N=\{y:|y|\le\mathcal{R}/\epsilon\}} d^3y\, \frac{\Lambda^{\mu\nu}(\tau - \epsilon|\vec{x}-\vec{y}|, y^k; \epsilon)}{|\vec{x}-\vec{y}|}, \quad (2.69)$$

$$h^{\mu\nu}_{P_N(F)} \equiv 4\int_{F=\{y:|y|>\mathcal{R}/\epsilon\}} d^3y\, \frac{\Lambda^{\mu\nu}(\tau - \epsilon|\vec{x}-\vec{y}|, y^k; \epsilon)}{|\vec{x}-\vec{y}|}. \quad (2.70)$$

$h^{\mu\nu}_{P_N(N)}$ is the near zone integral contribution to the near zone field and $h^{\mu\nu}_{P_N(F)}$ is the far zone integral contribution to the near zone field. The far zone contribution can be evaluated with the DIRE method developed in Pati and Will (2000). An explicit calculation shows that apparently there are far zone contributions to the near zone field at 3PN order. However these 3PN contributions are merely gauge terms. Pati and Will showed that the far zone contribution does not affect the equations of motion up to 3PN order inclusively and that the far zone contribution first appears at 4PN order. This result is consistent with the earlier result of Blanchet and Damour (1988) who used the multipolar-post-Minkowskian formalism. We follow the DIRE method and verify that the far zone contribution does not affect the equations of motion up to 3PN order in Appendix A. Henceforth we shall focus our attention on the near zone contribution $h^{\mu\nu}_{P_N(N)}$ and so we will ignore the far zone contribution in the following calculation of the field. As far as the homogeneous solution is concerned, we shall ignore it for simplicity. If we take random initial data for the field (Schutz, 1980) assumed to be of 1PN order (Futamase, 1983), it is irrelevant to the dynamics of the binary system up to the radiation reaction order (Futamase, 1983). Since we have assumed that the magnitude of the free data of the gravitational field on the initial hypersurface is 2.5PN order, we expect that the homogeneous solution does not affect the equations of motion up to 3PN order. We leave a full implementation of the initial value formulation for future work. It is worth noting that if we let τ become sufficiently large then the condition $h^{\mu\nu}_{\rm H}(\tau, x^i) = 0$ corresponds to a no-incoming-radiation condition at (Minkowskian) past null infinity [see for example Fock (1959)]. Equation (2.67) can be rewritten in the form of a Kirchhoff formula as

$$h^{\mu\nu}_{\rm H}(\tau, x^i) = \oint_{\partial C(\tau, x^i)} \frac{d\Omega_y}{4\pi} \left[\frac{\partial}{\partial \rho}(\rho h^{\mu\nu}(\tau', y^i)) + \frac{\partial}{\partial \tau'}(\rho h^{\mu\nu}(\tau', y^i)) \right] \Bigg|_{\tau'=0,\, \rho=|\vec{x}-\vec{y}|=\tau} . \tag{2.71}$$

Then the no-incoming-radiation condition at (Minkowskian) past null infinity,

$$\lim_{\substack{\tau=r,\\ r\to\infty}} \left[\frac{\partial}{\partial r}(r h^{\mu\nu}) + \frac{\partial}{\partial \tau}(r h^{\mu\nu}) \right] = 0, \tag{2.72}$$

is a sufficient condition to ensure that $h^{\mu\nu}_{\rm H}(\tau, x^i) = 0$ when τ tends to infinity.

2.10 Near zone and body zone contributions

The near zone contribution to the near zone field is given by

$$h^{\mu\nu}(\tau, x^i) = 4 \int_{N(\tau, x^k)} d^3y \, \frac{\Lambda^{\mu\nu}(\tau - \epsilon|\vec{x}-\vec{y}|, y^k; \epsilon)}{|\vec{x}-\vec{y}|}. \tag{2.73}$$

Henceforth we shall omit the subscript $P_N(N)$ on $h^{\mu\nu}$ for notational simplicity. We begin by making a retarded expansion of (2.73) and changing the integration region to a $\tau =$ constant spacelike hypersurface resulting in

$$h^{\mu\nu} = 4 \sum_{n=0} \frac{(-\epsilon)^n}{n!} \left(\frac{\partial}{\partial \tau} \right)^n \int_N d^3y \, |\vec{x}-\vec{y}|^{n-1} \Lambda^{\mu\nu}_N(\tau, y^k; \epsilon). \tag{2.74}$$

Note that the above integral depends on the arbitrary length $\mathcal{R}$ in general. Cancellation between the $\mathcal{R}$ dependent terms in the far zone contribution and those in the near zone contribution was shown, in Pati and Will (2000), to hold for all the post-Newtonian orders. In the following we shall omit terms with negative powers of $\mathcal{R}$ and let $\mathcal{R} \to 0$ whenever it gives a convergent result. On the other hand we shall retain terms having positive powers of $\mathcal{R}$, and logarithmic terms ($\ln \mathcal{R}$–terms), in order to confirm that the final result is independent of $\mathcal{R}$. Now we split the integral into two parts: a contribution from the body zone B_A, and a contribution from elsewhere (from N/B). Schematically we thus evaluate the following two types of integrals (omitting indices on $h^{\mu\nu}$ for convenience),

$$
\begin{aligned}
h &= h_B + h_{N/B}, \\
h_B &= \epsilon^6 \sum_{A=1,2} \int_{B_A} d^3\alpha_A \frac{f(\tau, \vec{z}_A + \epsilon^2\vec{\alpha}_A)}{|\vec{r}_A - \epsilon^2\vec{\alpha}_A|^{1-n}}, \\
h_{N/B} &= \int_{N/B} d^3y \frac{f(\tau, \vec{y})}{|\vec{x} - \vec{y}|^{1-n}},
\end{aligned}
\tag{2.75}
$$

where $\vec{r}_A \equiv \vec{x} - \vec{z}_A$. For the body zone contribution we make a multipole expansion using the scaling of the integrand, i.e. $\Lambda^{\mu\nu}$ in the body zone. For example the $n = 0$ contribution to (2.74) in the body zone, which we denote by $h^{\mu\nu}_{B\,n=0}$, is given by

$$
h^{\tau\tau}_{B\,n=0} = 4\epsilon^4 \sum_{A=1,2} \left(\frac{P^\tau_A}{r_A} + \epsilon^2 \frac{D^k_A r^k_A}{r^3_A} + \epsilon^4 \frac{3I^{\langle kl\rangle}_A r^k_A r^l_A}{2r^5_A} + \epsilon^6 \frac{5I^{\langle klm\rangle}_A r^k_A r^l_A r^m_A}{2r^7_A} \right)
+ O(\epsilon^{12}),
\tag{2.76}
$$

$$
h^{\tau i}_{B\,n=0} = 4\epsilon^4 \sum_{A=1,2} \left(\frac{P^i_A}{r_A} + \epsilon^2 \frac{J^{ki}_A r^k_A}{r^3_A} + \epsilon^4 \frac{3J^{kli}_A r^k_A r^l_A}{2r^5_A} \right) + O(\epsilon^{10}),
\tag{2.77}
$$

$$
h^{ij}_{B\,n=0} = 4\epsilon^2 \sum_{A=1,2} \left(\frac{Z^{ij}_A}{r_A} + \epsilon^2 \frac{Z^{kij}_A r^k_A}{r^3_A} + \epsilon^4 \frac{3Z^{\langle kl\rangle ij}_A r^k_A r^l_A}{2r^5_A} + \epsilon^6 \frac{5Z^{\langle klm\rangle ij}_A r^k_A r^l_A r^m_A}{2r^7_A} \right)
+ O(\epsilon^{10}).
\tag{2.78}
$$

Here the operator $\langle \ldots \rangle$ denotes a symmetric and tracefree (STF) operation on the indices in the brackets. The reader can consult Pati and Will (2000), Will and Wiseman (1996), and Thorne (1980) for some useful formulas for STF tensors. Also we define $r_A \equiv |\vec{r}_A|$. To derive the 3PN equations of motion, $h^{\tau\tau}$ up to $O(\epsilon^{10})$ and $h^{\tau i}$ as well as h^{ij} up to $O(\epsilon^8)$ are required. In the above equations we define the multipole moments as

$$
I^{K_l}_A \equiv \epsilon^2 \int_{B_A} d^3\alpha_A \, \Lambda^{\tau\tau}_N \, \alpha^{\underline{K_l}}_A,
\tag{2.79}
$$

$$
J^{K_l i}_A \equiv \epsilon^4 \int_{B_A} d^3\alpha_A \, \Lambda^{\tau \underline{i}}_N \, \alpha^{\underline{K_l}}_A,
\tag{2.80}
$$

$$
Z^{K_l ij}_A \equiv \epsilon^8 \int_{B_A} d^3\alpha_A \, \Lambda^{\underline{ij}}_N \, \alpha^{\underline{K_l}}_A,
\tag{2.81}
$$

where we have introduced a collective multi-index $I_l \equiv i_1 i_2 \ldots i_l$ and $\alpha^{\underline{I_l}}_A \equiv \alpha^{\underline{i_1}}_A \alpha^{\underline{i_2}}_A \ldots \alpha^{\underline{i_l}}_A$. Then $P^\tau_A \equiv I^{K_0}_A$, $D^{k_1}_A \equiv I^{K_1}_A$, and $P^{k_1}_A \equiv J^{K_1}_A$. We simply call P^μ_A the 4-momentum of the star A, P^i_A the momentum, and P^τ_A the energy. It is just for simplicity that we call P^μ_A a 4-momemtum. P^μ_A is not defined (general or even Lorentz) covariantly nor is it a volume integral of a tensor density of weight -1. The integrand $\Lambda^{\mu\nu}$ is a tensor density of weight -2. Later we will find that P^{μ}_{A} equals $\sqrt{-g}m_A u^{\mu}_{A}$ in a certain sense. P^{τ}_{A} is an energy of the star (modulo the contribution from the integrand $\chi^{\mu\nu\alpha\beta}{}_{,\alpha\beta}$) if the star considered is isolated, the space-time is asymptotically flat, and if we integrate over a spacelike hypersurface rather than integrate only over the body zone of the star (Landau and Lifshitz, 1975). In addition we call D^{k}_{A} the dipole moment of star A and I^{kl}_{A} the quadrupole moment of A. We can transform these moments into more convenient forms. By the conservation law (2.65) we have

$$\Lambda^{\tau i}_{N} = \left(\Lambda^{\tau\tau}_{N} y^{i}_{A}\right)_{,\tau} + \left(\Lambda^{\tau j}_{N} y^{i}_{A}\right)_{,j} + v^{i}_{A}\Lambda^{\tau\tau}_{N}, \tag{2.82}$$

$$\Lambda^{ij}_{N} = \left(\Lambda^{\tau(i}_{N} y^{j)}_{A}\right)_{,\tau} + \left(\Lambda^{k(i}_{N} y^{j)}_{A}\right)_{,k} + v^{(i}_{A}\Lambda^{j)\tau}_{N}, \tag{2.83}$$

where $v^{i}_{A} \equiv \dot{z}^{i}_{A}$, an overdot denotes a time derivative with respect to τ, and $\vec{y}_A \equiv \vec{y} - \vec{z}_A$. Using these equations and noticing that the body zone remains unchanged in the near zone coordinates, i.e. $\dot{R}_A = 0$, we have

$$P^{i}_{A} = P^{\tau}_{A} v^{i}_{A} + Q^{i}_{A} + \epsilon^{2}\frac{dD^{i}_{A}}{d\tau}, \tag{2.84}$$

$$J^{ij}_{A} = \frac{1}{2}\left(M^{ij}_{A} + \epsilon^{2}\frac{dI^{ij}_{A}}{d\tau}\right) + v^{(i}_{A}D^{j)}_{A} + \frac{1}{2}\epsilon^{-2}Q^{ij}_{A}, \tag{2.85}$$

$$\begin{aligned} Z^{ij}_{A} &= \epsilon^{2}P^{\tau}_{A}v^{i}_{A}v^{j}_{A} + \frac{1}{2}\epsilon^{6}\frac{d^{2}I^{ij}_{A}}{d\tau^{2}} + 2\epsilon^{4}v^{(i}_{A}\frac{dD^{j)}_{A}}{d\tau} + \epsilon^{4}\frac{dv^{(i}_{A}}{d\tau}D^{j)}_{A} + \epsilon^{2}Q^{(i}_{A}v^{j)}_{A} \\ &+ \epsilon^{2}R^{(ij)}_{A} + \frac{1}{2}\epsilon^{2}\frac{dQ^{ij}_{A}}{d\tau}, \end{aligned} \tag{2.86}$$

$$Z^{kij}_{A} = \frac{3}{2}A^{kij}_{A} - A^{(ij)k}_{A}, \tag{2.87}$$

where

$$M^{ij}_{A} \equiv 2\epsilon^{4}\int_{B_A} d^{3}\alpha_{A}\, \alpha^{[i}_{A}\Lambda^{j]\tau}_{N}, \tag{2.88}$$

$$Q^{K_l i}_{A} \equiv \epsilon^{-4}\oint_{\partial B_A} dS_k \left(\Lambda^{\tau k}_{N} - v^{k}_{A}\Lambda^{\tau\tau}_{N}\right) y^{K_l}_{A} y^{i}_{A}, \tag{2.89}$$

$$R^{K_l ij}_{A} \equiv \epsilon^{-4}\oint_{\partial B_A} dS_k \left(\Lambda^{kj}_{N} - v^{k}_{A}\Lambda^{\tau j}_{N}\right) y^{K_l}_{A} y^{i}_{A}, \tag{2.90}$$

and

$$A^{kij}_{A} \equiv \epsilon^{2}J^{k(i}_{A}v^{j)}_{A} + \epsilon^{2}v^{k}_{A}J^{(ij)}_{A} + R^{k(ij)}_{A} + \epsilon^{4}\frac{dJ^{k(ij)}_{A}}{d\tau}. \tag{2.91}$$

The operators [ ] and ( ) attached to the indices denote anti-symmetrization and symmetrization respectively. M^{ij}_{A} is the spin of the star A and (2.84) is the momentum–velocity relation. Thus our momentum–velocity relation is a direct analog of the Newtonian momentum–velocity relation. In general we have

$$J_A^{K_l i} = J_A^{(K_l i)} + \frac{2l}{l+1} J_A^{(K_{l-1}[k_l)i]}, \tag{2.92}$$

$$Z_A^{K_l ij} = \frac{1}{2} Z_A^{(K_l i)j} + \frac{l}{l+1} Z_A^{(K_{l-1}[k_l)i]j} + \frac{1}{2} Z_A^{(K_l j)i}$$
$$+ \frac{l}{l+1} Z_A^{(K_{l-1}[k_l)j]i}, \tag{2.93}$$

and

$$J_A^{(K_l i)} = \frac{1}{l+1} \epsilon^2 \frac{dI_A^{K_l i}}{d\tau} + v_A^{(i} I_A^{K_l)} + \frac{1}{l+1} \epsilon^{-2l} Q_A^{K_l i}, \tag{2.94}$$

$$Z_A^{(K_l i)j} + Z_A^{(K_l j)i} = \epsilon^2 v_A^{(i} J_A^{K_l)j} + \epsilon^2 v_A^{(j} J_A^{K_l)i} + \frac{2}{l+1} \epsilon^4 \frac{dJ_A^{K_l (ij)}}{d\tau}$$
$$+ \epsilon^{-2l+2} R_A^{K_l (ij)}, \tag{2.95}$$

where l is the number of indices in the multi-index K_l.

From the above equations, especially (2.86), we find that the body zone contributions, $h^{\mu\nu}_{B\,n=0}$, are of order $O(\epsilon^4)$. Note that if we cannot or do not assume the (almost) stationarity of the initial data for the stars then instead of (2.86) we have

$$Z_A^{ij} = \epsilon^2 P_A^\tau v_A^i v_A^j + \frac{1}{2} \frac{d^2 I_A^{ij}}{d\eta^2} + \dots, \tag{2.96}$$

where we have used the dynamical time η introduced earlier. In this case the lowest-order metric differs from the Newtonian form. From our (almost) stationarity assumption the remaining motion inside a star, apart from the spinning motion, is caused only by the tidal effect due to the companion star and from (2.86); it appears at 3PN order (Futamase, 1987). To obtain the lowest-order $h^{\mu\nu}_{B\,n=0}$ we have to evaluate the surface integrals $Q_A^{K_l i}$ and $R_A^{K_l ij}$. Generally in $h^{\mu\nu}_{B\,n=0}$ the moments $J_A^{K_l i}$ and $Z_A^{K_l ij}$ appear formally at orders ϵ^{2l+4} and ϵ^{2l+2}. Thus $Q_A^{K_l i}$ and $R_A^{K_l ij}$ arise as

$$h^{\tau i}_{B\,n=0} \sim \dots + \epsilon^4 \frac{r_A^{K_l} Q_A^{\langle K_l \rangle i}}{r_A^{2l+1}} + \dots, \tag{2.97}$$

$$h^{ij}_{B\,n=0} \sim \dots + \epsilon^4 \frac{r_A^{K_l} R_A^{\langle K_l \rangle (ij)}}{r_A^{2l+1}} + \dots, \tag{2.98}$$

where we have omitted irrelevant terms and numerical coefficients. Thus one may expect that $Q_A^{K_l i}$ and $R_A^{K_l ij}$ appear at order ϵ^4 for any l and we have to calculate an infinite number of moments. In fact this is not the case and it has been shown in Itoh (2004) that only the $l = 0, 1$ terms of $R_A^{K_l ij}$ contribute to the 3PN equations of motion. The important thing here is that $\epsilon^4 Q_A^{K_l i}$ and $\epsilon^4 R_A^{K_l ij}$ are at most $O(\epsilon^4)$ in

$h^{\mu\nu}_{Bn=0}$. Since the order of $h^{\mu\nu}_{Bn}(n \geq 1)$ is higher than that of $h^{\mu\nu}_{Bn=0}$ we conclude that $h^{\mu\nu}_B = O(\epsilon^4)$.

Finally, as far as the N/B contribution is concerned, since the integrand $\Lambda^{\mu\nu}_N = -gt^{\mu\nu}_{\rm LL} + \chi^{\mu\nu\alpha\beta}{}_{,\alpha\beta}$ is at least quadratic in the small deviation field $h^{\mu\nu}$, we make the post-Newtonian expansion in the integrand. Then with the help of (super-)potentials $g(\vec{x})$ which satisfy $\Delta g(\vec{x}) = f(\vec{x})$, Δ denoting the Laplacian, we have for each integral [see for example the $n = 0$ term in (2.75)]

$$\int_{N/B} d^3y \frac{f(\vec{y})}{|\vec{x}-\vec{y}|} = \oint_{\partial(N/B)} dS_k \left[\frac{1}{|\vec{x}-\vec{y}|}\frac{\partial g(\vec{y})}{\partial y^k} - g(\vec{y})\frac{\partial}{\partial y^k}\left(\frac{1}{|\vec{x}-\vec{y}|}\right)\right] -4\pi g(\vec{x}). \tag{2.99}$$

Equation (2.99) can be derived without using Dirac delta functions [see Appendix B of Itoh (2004)]. For the $n \geq 1$ terms in (2.75) potentials are used many times to convert all the volume integrals into surface integrals and "$-4\pi g(\vec{x})$" terms. We note that when solving a Poisson equation $\Delta g(\vec{x}) = f(\vec{x})$ a particular solution suffices for our purposes. By virtue of the surface integral term in (2.99) it is not necessary to take into account a homogeneous solution of Poisson's equation. Finding the super-potentials is one of the most formidable tasks to be carried out, especially when we proceed to high post-Newtonian orders. Fortunately up to 2.5PN order all the required super-potentials are available. At 3PN order there appear many integrands for which the required super-potentials are not available. To obtain the 3PN equations of motion we have devised an alternative method which is similar to the method employed by Blanchet and Faye (2001*a*). The details of the method will be explained later. Now the lowest-order integrands can be evaluated with the body zone contribution $h^{\mu\nu}_B$, and since $h^{\mu\nu}_B$ is $O(\epsilon^4)$, we find

$$\Lambda^{\tau\tau}_N = O(\epsilon^6), \tag{2.100}$$

$$\Lambda^{\tau i}_N = O(\epsilon^6), \tag{2.101}$$

$$\begin{aligned}\Lambda^{ij}_N &= (-g)t^{ij}_{\rm LL} + O(\epsilon^8), \\ &= \epsilon^4 \frac{1}{64\pi}\left(\delta^i{}_k\delta^j{}_l - \frac{1}{2}\delta^{ij}\delta_{kl}\right){}_4h^{\tau\tau,k}_B{}_4h^{\tau\tau,l}_B + O(\epsilon^5),\end{aligned} \tag{2.102}$$

where we have expanded $h^{\mu\nu}_B$ in an ϵ series,

$$h^{\mu\nu}_B = \sum_{n=0} \epsilon^{4+n}{}_nh^{\mu\nu}_B. \tag{2.103}$$

Similarly in the sequel we will expand $h^{\mu\nu}$ in an ϵ series. From these equations we find that $h^{\mu\nu}$, the deviation field in N/B, is $O(\epsilon^4)$. In contrast it should be noted that in the body zone $h^{\mu\nu}$ is assumed to be of order unity and with our method we cannot calculate $h^{\mu\nu}$ there explicitly. To obtain $h^{\mu\nu}$ in the body zone we have to know the internal structure of the star.

2.11 Lorentz contraction and multipole moments

In equations (2.79, 2.80, 2.81) we have defined multipole moments of a star. The definitions of those multipole moments are operational ones and are not necessarily equal to "intrinsic" multipole moments of the star. This is clear, for example, if we remember that a moving ball has spurious multipole moments due to Lorentz contraction. We have adopted the generalized Fermi normal coordinates (GFC) (Ashby and Bertotti, 1986) as the star's reference coordinates to address this problem. A question specific to our formalism is whether the differences between the multipole moments defined in (2.79, 2.80, 2.81) and the multipole moments in GFC give purely monopole terms. If so we would of course have to take into account such terms in the field to compute the equations of motion for two intrinsically spherical star binaries. This problem is addressed in Appendix C of Itoh (2004) and the differences are mainly attributable to the shape of the body zone. The body zone B_A which is spherical in the near zone coordinates (NZC) is not spherical in the GFC mainly because of a kinematic effect (Lorentz contraction). To derive the 3PN equations of motion it turns out that it is sufficient to compute the difference in the STF quadrupole moment up to 1PN order. The result is

$$\begin{aligned}
\delta I_A^{\langle ij\rangle} &\equiv I_{A,\mathrm{NZC}}^{\langle ij\rangle} - I_{A,\mathrm{GFC}}^{\langle ij\rangle}, \\
&= \epsilon^{-6}\frac{1}{2}v_A^k v_A^l \oint_{\partial B_A} dS_k y_A^l\, y_A^{\langle i} y_A^{j\rangle} \Lambda^{\tau\tau} + O(\epsilon^3), \\
&= -\epsilon^2 \frac{4m_A^3}{5} v_A^{\langle i} v_A^{j\rangle} + O(\epsilon^3), \qquad (2.104)
\end{aligned}$$

where $I_{A,\mathrm{NZC}}^{ij} \equiv I_A^{ij}$. $I_{A,\mathrm{GFC}}^{ij}$ are the quadrupole moments defined in the generalized Fermi normal coordinates. Note that the difference $\delta I_A^{\langle ij\rangle}$ is expressed in the form of a surface integral. It is obvious from (2.104) that this difference appears even if the companion star does not exist. We note that we could derive the 3PN metric for an isolated star A moving at a constant velocity, using the method explained in this chapter, by simply letting the mass of the companion star be zero. Actually $\delta I_A^{\langle ij\rangle}$ above is a necessary term which makes the 3PN metric obtained equal to the Schwarzschild metric boosted to the constant velocity $\vec{v}_A$ in harmonic coordinates.

The reason why the influence of a body's Lorentz contraction appears from 3PN order is as follows. The Lorentz contraction arises as an apparent deformation of the body, manifested as an apparent quadrupole moment, to leading order in the frame in which the body is moving. The body's (real) quadrupole affects the field from 2PN order through [see (2.76)]

$$h_{\mathrm{quad}}^{\tau\tau} = \epsilon^8 \sum_A \frac{3I_A^{\langle ij\rangle} r_A^i r_A^j}{2r_A^5}. \qquad (2.105)$$

The radius of a compact body A is of the order of its mass m_A, and the Lorentz contraction deforms the body so that the radius of the body changes by the amount

$\epsilon^2 m_A v_A^2 + O(\epsilon^4)$. So the apparent quadrupole moment due to Lorentz contraction is of the order of

$$I^{\langle ij \rangle}_{A\,\mathrm{apparent}} \sim \epsilon^2 m_A^3 v_A^2. \tag{2.106}$$

This field is the 3PN field and thus the Lorentz contraction of a body affects the equations of motion starting from 3PN order.

2.12 General form of the equations of motion

From the definition of the 4-momentum,

$$P_A^\mu(\tau) = \epsilon^2 \int_{B_A} d^3\alpha_A \Lambda_N^{\tau\mu}, \tag{2.107}$$

and the conservation law (2.65), we have the evolution equations for the 4-momentum:

$$\frac{dP_A^\mu}{d\tau} = -\epsilon^{-4} \oint_{\partial B_A} dS_k\, \Lambda_N^{k\mu} + \epsilon^{-4} v_A^k \oint_{\partial B_A} dS_k\, \Lambda_N^{\tau\mu}. \tag{2.108}$$

Here we have used the fact that the size and the shape of the body zone are defined to be fixed (in the near zone coordinates), while the center of the body zone moves at the velocity of the star's representative point. Substituting the momentum–velocity relation (2.84) into the spatial components of (2.108) we obtain the general form of equations of motion for star A as

$$\begin{aligned} P_A^\tau \frac{dv_A^i}{d\tau} = &-\epsilon^{-4} \oint_{\partial B_A} dS_k\, \Lambda_N^{ki} + \epsilon^{-4} v_A^k \oint_{\partial B_A} dS_k\, \Lambda_N^{\tau i} \\ &+ \epsilon^{-4} v_A^i \left(\oint_{\partial B_A} dS_k\, \Lambda_N^{k\tau} - v_A^k \oint_{\partial B_A} dS_k\, \Lambda_N^{\tau\tau} \right) \\ &- \frac{dQ_A^i}{d\tau} - \epsilon^2 \frac{d^2 D_A^i}{d\tau^2}. \end{aligned} \tag{2.109}$$

All of the terms on the right-hand side of (2.109), except the dipole moment, are expressed as surface integrals. We can specify the value of D_A^i freely to determine the representative point $z_A^i(\tau)$ of star A. Up to 2.5PN order we take $D_A^i = 0$ and simply call z_A^i the center of mass of star A. Note that in order to obtain the spin–orbit coupling force in the same form as in previous works (Damour, 1982; Kidder, 1995; Thorne and Hartle, 1985) we have to make another choice for z_A^i [see Itoh *et al.* (2000) and Appendix B]. At 3PN order, in the sequel, yet another choice is made for the dipole moment D_A^i. In (2.109) the factor P_A^τ rather than the mass of star A appears. Hence we must derive a relation between the mass and P_A^τ. We shall derive that relation by solving the temporal component of the evolution equations (2.108) functionally. Then since all terms in the equations are expressed in terms of surface integrals, with the exception of D_A^i to be specified, we can derive the equations of motion for a strongly self-gravitating star using the post-Newtonian approximation.

2.13 Remarks on the arbitrary constant R_A

Since we have introduced the arbitrary body zone size R_A the constant R_A seems to appear in the metric, the multipole moments of the stars, and the equations of motion. This is on account of the splitting of the deviation field into two parts (the contributions from B and N/B), the definition of the moments and the surface integrals that we evaluate to derive the equations of motion. Consequently these issues need to be addressed.

The B and N/B contributions to the field depend on the body zone boundary ϵR_A. But $h^{\mu\nu}$ itself does not depend on ϵR_A. Thus it is natural to expect that there are renormalized multipole moments which are independent of R_A since we use nonsingular matter sources. This renormalization would absorb the ϵR_A dependence occurring in the computation of the N/B field (see Section 2.15 below for an example of such a renormalization). One possible practical obstacle to this expectation might be the $\ln(\epsilon R_A)$ dependence of multipole moments. Although at 3PN order there appear such logarithmic terms, it has been found that they can be removed by a choice of dipole moment D_A^i of the star. Though we will use the same symbol for the moments henceforth as before, for notational simplicity, it should be understood that they are the renormalized ones.

Since we compute integrals over the body zone boundary in general the resulting equations of motion would appear to depend on the size of the body zone boundary ϵR_A. Actually this is not the case. In the derivation of (2.109), had we not used the conservation law (2.65) until the final step, we would have obtained

$$
\begin{aligned}
P_A^\tau \frac{dv_A^i}{d\tau} &= -\epsilon^{-4} \oint_{\partial B_A} dS_k\, \Lambda_N^{ki} + \epsilon^{-4} v_A^k \oint_{\partial B_A} dS_k\, \Lambda_N^{\tau i} \\
&+ \epsilon^{-4} v_A^i \left(\oint_{\partial B_A} dS_k\, \Lambda_N^{k\tau} - v_A^k \oint_{\partial B_A} dS_k\, \Lambda_N^{\tau\tau} \right) \\
&- \frac{dQ_A^i}{d\tau} - \epsilon^2 \frac{d^2 D_A^i}{d\tau^2} \\
&+ \epsilon^{-4} \int_{B_A} d^3y\, \Lambda_{N,\nu}^{i\nu} - \epsilon^{-4} v_A^i \int_{B_A} d^3y\, \Lambda_{N,\nu}^{\tau\nu} \\
&+ \epsilon^{-4} \frac{d}{d\tau} \left(\int_{B_A} d^3y\, \Lambda_{N,\nu}^{\tau\nu} y_A^i \right).
\end{aligned}
\tag{2.110}
$$

Since the conservation law is satisfied irrespective of the value of R_A the right-hand side of this equation is zero for any R_A. Hence the equations of motion (2.109) do not depend on R_A [a similar argument can be found in Einstein *et al.* (1938)]. In a similar fashion one sees that the momentum–velocity relation (2.84) does not depend on R_A. In the sequel we shall explicitly show the independence of the field and the equations of motion on R_A by checking the cancellation among the R_A dependent terms up to 0.5PN order.

2.14 The Newtonian and 1PN equations of motion

We first derive the Newtonian equations of motion and the 1PN correction to the acceleration. The derivation of the 1PN equations of motion illustrates some essential features of our formalism and shows how it works. Hence we shall give a detailed explanation of the derivation although the calculation is elementary and the resulting equations of motion are well known.

We first derive the Newtonian mass–energy relation. From (2.100, 2.101) and the time component of (2.108) we have

$$\frac{dP_A^\tau}{d\tau} = O(\epsilon^2). \tag{2.111}$$

We define the mass of star A as the constant of integration of this equation,

$$m_A \equiv \lim_{\epsilon\to 0} P_A^\tau. \tag{2.112}$$

Here m_A is the ADM mass that star A would have if it were isolated. We have taken the ϵ tending to zero limit in (2.112) to ensure that the mass defined above does not include the effect of the companion star and/or the orbital motion of the star itself. Thus we ensure that this mass is the constant of integration of (2.111). Next we obtain the lowest-order component $h^{\tau\tau}$:

$${}_4h^{\tau\tau} = {}_4h_B^{\tau\tau} = 4\sum_{A=1,2}\frac{m_A}{r_A}. \tag{2.113}$$

Using (2.89) with (2.100) and (2.101) we obtain $Q_A^i = 0$ at the lowest order. Thus we have the Newtonian momentum–velocity relation $P_A^i = m_A v_A^i + O(\epsilon^2)$ from (2.84) (setting $D_A^i - 0$). Now substituting (2.102) with the lowest-order $h^{\tau\tau}$ into the general form of the equations of motion (2.109) and using the Newtonian momentum–velocity relation we obtain for star 1:

$$\begin{aligned}
m_1\frac{dv_1^i}{d\tau} &= -\oint_{\partial B_1} dS_{k(4)}\left[(-g)t_{\rm LL}^{ik}\right],\\
&= -\frac{1}{4\pi}\left(\delta_l^i\delta_m^k - \frac{1}{2}\delta^{ik}\delta_{lm}\right)\sum_{A=1,2}\sum_{B=1,2}\oint_{\partial B_1} dS_k\,\frac{m_A m_B y_A^l y_B^m}{y_A^3 y_B^3},\\
&= -\frac{1}{4\pi}\left(\delta_l^i\delta_m^k - \frac{1}{2}\delta^{ik}\delta_{lm}\right)\oint d\Omega_{\mathbf{n_1}}\, n_1^k\\
&\quad\times\left(\frac{m_1^2 n_1^l n_1^m}{(\epsilon R_1)^2} + \frac{2m_1 m_2 n_A^{(l}\left(r_{12}^{m)} + \epsilon R_1 n_1^{m)}\right)}{|\vec{r}_{12} + \epsilon R_1\vec{n}_1|^3}\right.\\
&\qquad\left. + \frac{(\epsilon R_1)^2 m_2^2\left(r_{12}^l + \epsilon R_1 n_1^l\right)\left(r_{12}^m + \epsilon R_1 n_1^m\right)}{|\vec{r}_{12} + \epsilon R_1\vec{n}_1|^3}\right),
\end{aligned} \tag{2.114}$$

where y^i_A is the integration variable and $\vec{y}_A = \vec{y} - \vec{z}_A$. Here $\vec{n}_1$ is the unit 3-vector in the direction of $\vec{z}_1$. Every spatial 3-vector is treated as a Euclidean 3-vector. In Cartesian coordinates the norm is evaluated with the Kronecker delta. Also $\vec{r}_{12} \equiv \vec{z}_1 - \vec{z}_2$ and $n^i_{12} \equiv r^i_{12}/r_{12}$. We have used $\vec{r}_1 = \epsilon R_1 \vec{n}_1$ and $\vec{r}_2 = \vec{r}_{12} + \epsilon R_1 \vec{n}_1$. In the above equation, on account of the integral with respect to the angles specifying the directions of $\vec{n}_1$, the first term (which would be singular when the ϵ tending to zero limit is taken) vanishes. The third term vanishes by letting ϵ go to zero. Only the second term survives and gives the Newtonian equations of motion as expected. This completes the Newtonian order calculations.

Moving now to the 1PN order we first require ${}_6h^{\tau\tau}$ and ${}_4h^{\mu\nu}$. The $n = 1$ term in the retardation expansion series of $h^{\tau\tau}$ [obtained from (2.74)] gives no contribution at the 1PN order on account of the constancy of mass m_A. Thus ${}_5h^{\tau\tau} = 0$. Now we obtain ${}_4h^{\tau i}$ from the Newtonian momentum–velocity relation as

$$ {}_4h^{\tau i} = 4 \sum_{A=1,2} \frac{m_A v^i_A}{r_A}. \tag{2.115} $$

Evaluating the surface integrals in the evolution equation for P^τ_A at 1PN order results, for star 1, in

$$ \frac{dP^\tau_1}{d\tau} = \epsilon^2 m_1 \frac{d}{d\tau}\left(\frac{1}{2}v_1^2 + \frac{3m_2}{r_{12}}\right) + O(\epsilon^3), \tag{2.116} $$

where we have used the Newtonian equations of motion. From this equation we have the mass–energy relation at 1PN order

$$ P^\tau_1 = m_1 \left[1 + \epsilon^2 \left(\frac{1}{2}v_1^2 + \frac{3m_2}{r_{12}}\right)\right] + O(\epsilon^3). \tag{2.117} $$

Now we must calculate the 1PN order Q^i_A. The result is $Q^i_A = \epsilon^2 m_A^2 v^i_A/(6\epsilon R_A)$. Since Q^i_A depends on R_A we can ignore it. As a result we obtain the momentum–velocity relation at 1PN order in the form $P^i_A = P^\tau_A v^i_A + O(\epsilon^3)$ from (2.84). To obtain ${}_4h^{ij}$ we first calculate the surface integrals Q^{ij}_A, R^{ij}_A, and R^{kij}_A from (2.89) and (2.90). We find that they depend on R_A and hence we can ignore them and arrive at

$$ h^{ij}_B = 4\epsilon^4 \sum_{A=1,2} \frac{m_A v^i_A v^j_A}{r_A} + O(\epsilon^5). \tag{2.118} $$

To obtain ${}_6h^{\tau\tau}$ and ${}_4h^{ij}$, we have to evaluate non-compact support integrals for ${}_6h^{\tau\tau}_{N/B}$ and ${}_4h^{ij}_{N/B}$, and the $n = 2$ term in (2.74) for $h^{\tau\tau}$. The result is

$$ \begin{aligned} h^{\tau\tau} &= 4\epsilon^4 \sum_{A=1,2} \frac{P^\tau_A}{r_A} + 4\epsilon^6 \int_{N/B} \frac{d^3y}{|\vec{x} - \vec{y}|} {}_6\left[(-g)t^{\tau\tau}_{\rm LL}\right] \\ &\quad + 2\epsilon^6 \frac{\partial^2}{\partial\tau^2} \sum_{A=1,2} P^\tau_A r_A, \end{aligned} \tag{2.119} $$

$$ h^{ij} = 4\epsilon^4 \sum_{A=1,2} \frac{m_A v^i_A v^j_A}{r_A} + 4\epsilon^4 \int_{N/B} \frac{d^3y}{|\vec{x} - \vec{y}|} {}_4\left[(-g)t^{ij}_{\rm LL}\right], \tag{2.120} $$

with

$$
{}_6\left[(-16\pi g)t^{\tau\tau}_{\rm LL}\right] = -\frac{7}{8}{}_4h^{\tau\tau,k}{}_4h^{\tau\tau}{}_{,k}, \tag{2.121}
$$

$$
{}_4\left[(-16\pi g)t^{ij}_{\rm LL}\right] = \frac{1}{4}\left(\delta^i{}_k\delta^i{}_k - \frac{1}{2}\delta^{ij}\delta_{kl}\right){}_4h^{\tau\tau,k}{}_4h^{\tau\tau,l}. \tag{2.122}
$$

Evaluation of the twice retardation expansion term [the last term in (2.119)] is straightforward. For non-compact support integrals it is sufficient to consider the following integral:

$$
\begin{aligned}
\int_{N/B}\frac{d^3y}{16\pi|\vec{x}-\vec{y}|}{}_4h^{\tau\tau,i}{}_4h^{\tau\tau,j} &= \frac{1}{\pi}\sum_{A=1,2}P_A^{\tau 2}\int_{N/B}\frac{d^3y}{|\vec{x}-\vec{y}|}\frac{r_A^ir_A^j}{r_A^6} \\
&\quad + \frac{2}{\pi}P_1^\tau P_2^\tau\int_{N/B}\frac{d^3y}{|\vec{x}-\vec{y}|}\frac{r_1^{(i}r_2^{j)}}{r_1^3r_2^3}.
\end{aligned} \tag{2.123}
$$

To evaluate the above Poisson integrals we use (2.99) and thus we need to find super-potentials for the integrands. For this purpose it is convenient to transform the tensorial integrands into scalar integrands with derivative operators using

$$
\frac{r_A^ir_A^j}{r_A^6} = \frac{1}{8}\left(\delta^i{}_k\delta^j{}_l + \delta^{ij}\delta_{kl}\right)\frac{\partial^2}{\partial y^k\partial y^l}\left(\frac{1}{r_A^2}\right), \tag{2.124}
$$

$$
\frac{r_1^ir_2^j}{r_1^3r_2^3} = \frac{\partial}{\partial z_1^i}\frac{\partial}{\partial z_2^i}\left(\frac{1}{r_1r_2}\right). \tag{2.125}
$$

It is relatively easy to find the super-potentials for these scalars. The results are $\Delta\ln r_1 = 1/r_1^2$ and $\Delta\ln S = 1/(r_1r_2)$, where $S = r_1 + r_2 + r_{12}$ (Fock, 1959). Equation (2.99) for each integrand is now obtained from

$$
\frac{1}{4\pi}\int_{N/B}\frac{d^3y}{|\vec{x}-\vec{y}|}\frac{r_A^ir_A^j}{r_A^6} = -\frac{\delta^{ij}}{4r_A^2} + \frac{r_A^ir_A^j}{4r_A^4} + \frac{\delta^{ij}}{3\epsilon R_1r_1} + O(\epsilon R_A), \tag{2.126}
$$

and

$$
\begin{aligned}
\frac{1}{4\pi}\int_{N/B}\frac{d^3y}{|\vec{x}-\vec{y}|}\frac{r_1^ir_2^j}{r_1^3r_2^3} &= \frac{\delta^{ij}}{r_{12}S} + \frac{r_1^ir_{12}^j}{r_1r_{12}S^2} - \frac{r_{12}^ir_{12}^j}{r_{12}^2S^2} - \frac{r_{12}^ir_{12}^j}{r_{12}^3S} + \frac{r_1^ir_{12}^j}{r_1r_2S^2} \\
&\quad - \frac{r_{12}^ir_2^j}{r_{12}r_2S^2} + O(\epsilon R_A).
\end{aligned} \tag{2.127}
$$

The second to last term of (2.126) and the terms abbreviated as $O(\epsilon R_A)$ in the above two equations arise from the surface integrals in (2.99). Since they depend on ϵR_A we ignore them. Substituting (2.126) and (2.127) back into (2.123) we can compute the non-compact support integrals. Using the above results and the Newtonian equations of motion for the twice retardation expansion term, we finally obtain

$$
\begin{aligned}
h^{\tau\tau} = 4\epsilon^4 \sum_{A=1,2} \frac{P_A^\tau}{r_A} \\
+ \epsilon^6 \Bigg[-2 \sum_{A=1,2} \frac{m_A}{r_A} \{(\vec{n}_A \cdot \vec{v}_A)^2 - v_A^2\} + 2\frac{m_1 m_2}{r_{12}^2} \vec{n}_{12} \cdot (\vec{n}_1 - \vec{n}_2) \\
+ 7 \sum_{A=1,2} \frac{m_A^2}{r_A^2} + 14\frac{m_1 m_2}{r_1 r_2} - 14\frac{m_1 m_2}{r_{12}} \sum_{A=1,2} \frac{1}{r_A} \Bigg] \\
+O(\epsilon^7),
\end{aligned}
\tag{2.128}
$$

$$
\begin{aligned}
h^{ij} = 4\epsilon^4 \sum_{A=1,2} \frac{m_A v_A^i v_A^j}{r_A} \\
+ \epsilon^4 \Bigg[\sum_{A=1,2} \frac{m_A^2}{r_A^2} n_A^i n_A^j - \frac{8 m_1 m_2}{r_{12} S} n_{12}^i n_{12}^j \\
- 8 \left(\delta^i{}_k \delta^j{}_l - \frac{1}{2}\delta^{ij}\delta_{kl} \right) \frac{m_1 m_2}{S^2} (\vec{n}_{12} - \vec{n}_1)^{(k} (\vec{n}_{12} + \vec{n}_2)^{l)} \Bigg] \\
+O(\epsilon^5),
\end{aligned}
\tag{2.129}
$$

where $\vec{n}_A \equiv \vec{r}_A / r_A$. Evaluating the surface integrals in (2.109) as in the Newtonian case we obtain the 1PN equations of motion,

$$
\begin{aligned}
m_1 \frac{dv_1^i}{d\tau} = -\frac{m_1 m_2}{r_{12}^2} n_{12}^i \\
+ \epsilon^2 \frac{m_1 m_2}{r_{12}^2} \Bigg[n_{12}^i \Bigg(-v_1^2 - 2v_2^2 + \frac{3}{2}(\vec{n}_{12} \cdot \vec{v}_2)^2 + 4(\vec{v}_1 \cdot \vec{v}_2) \\
+ \frac{5 m_1}{r_{12}} + \frac{4 m_2}{r_{12}} \Bigg) \\
+ V^i \left(4(\vec{n}_{12} \cdot \vec{v}_1) - 3(\vec{n}_{12} \cdot \vec{v}_2)\right) \Bigg],
\end{aligned}
\tag{2.130}
$$

where we define the relative velocity as $\vec{V} \equiv \vec{v}_1 - \vec{v}_2$ and we have used the Newtonian equations of motion as well as (2.117).

The general pattern of our procedure that emerges is as follows. With the nPN-order equations of motion and $h^{\mu\nu}$ in hand, we first derive the $(n+1)$PN evolution equation for P_A^τ. Then we solve it functionally and obtain the mass–energy relation at $(n+1)$PN order. Next we calculate Q_A^i at $(n+1)$PN order and derive the momentum–velocity relation at $(n+1)$PN order. We then calculate $Q_A^{K_l i}$ and $R_A^{K_l ij}$. With the $(n+1)$PN mass–energy relation, the $(n+1)$PN momentum–velocity relation, $Q_A^{K_l i}$, and $R_A^{K_l ij}$, we next derive the $(n+1)$PN deviation field $h^{\mu\nu}$. Finally, we evaluate the surface integrals which appear on the right-hand side of (2.109) and obtain the $(n+1)$PN equations of motion. In the above calculations we use the nPN-order equations of motion to reduce the order of the equations of motion whenever an acceleration appears on the right in the resulting equations of motion. For instance when we meet $\epsilon^2 dv_1^i/d\tau$ in the right-hand side of the equations of motion and we have to evaluate this up to ϵ^2 then, using the Newtonian equations of motion, we replace it by $-\epsilon^2 m_2 r_{12}^i/r_{12}^3$. We shall derive the 3PN equations of motion in the next chapter following this procedure.

2.15 A comment on body zone boundary dependent terms

We have consistently discarded the body zone boundary dependent terms in $h^{\mu\nu}$ since we expect that they cancel out between the body zone B contribution and the N/B contribution. Before moving on to the higher-order calculations it is helpful to see that such a cancellation really occurs in the field and consequently in the equations of motion up to 0.5PN order.

First we show the independence of the body zone radius R_A in the 0.5PN field. Returning to the derivation of the 1PN $h^{\tau\tau}$, taking care to exhibit the R_A dependence [see (2.126)], we get

$$\begin{aligned} h^{\tau\tau} &= 4\epsilon^4 \sum_{A=1,2} \frac{P_A^\tau}{r_A} + 4\epsilon^6 \int_{C/B} d^3y \, \frac{{}_6\Lambda_N^{\tau\tau}(\tau, y^k)}{|\vec{x}-\vec{y}|} + 2\epsilon^6 \frac{\partial^2}{\partial\tau^2} \sum_{A=1,2} P_A^\tau r_A + O(\epsilon^7), \\ &= 4\epsilon^4 \sum_{A=1,2} \frac{1}{r_A}\left(P_A^\tau - \epsilon^2 \frac{7m_A^2}{2\epsilon R_A}\right) + O(\epsilon^6). \end{aligned} \tag{2.131}$$

Now we split P_A^τ according to $P_A^\tau = \bar{P}_A^\tau + \tilde{P}_A^\tau$ where $\bar{P}_A^\tau$ does not depend on R_A while $\tilde{P}_A^\tau$ does. In order to evaluate the R_A dependent part of P_A^τ, $\tilde{P}_A^\tau$, we use the fact that the integral of $\Lambda_N^{\tau\tau}$ over the near zone does not depend on the size of the body zone, at the same time noting that P_A^τ is defined as the volume integral of $\Lambda_N^{\tau\tau}$ over B_A. From the relevant expression for the pseudotensor we obtain at the lowest order

$$\epsilon^2 \int_{N/B} d^3\alpha \, \epsilon^6 {}_6\Lambda_N^{\tau\tau} = \epsilon^2 \int_{N/B} d^3y {}_6\Lambda_N^{\tau\tau} = -\epsilon^2 \frac{7}{2} \sum_{A=1,2} \frac{m_A^2}{\epsilon R_A} + \mathcal{I}(R_A), \tag{2.132}$$

where $\mathcal{I}(R_A)$ stands for terms independent of R_A or terms having positive powers of R_A. Hence we find that $\tilde{P}_A^\tau = \epsilon^2 7m_A^2/(2\epsilon R_A) + O(\epsilon^2)$. The above equation shows us that the 0.5PN field is independent of R_A and therefore correctly expressed by the R_A independent energy $\bar{P}_A^\tau$ or mass up to 0.5PN order.

In similar fashion we can split P^i_A according to $P^i_A = \bar{P}^i_A + \tilde{P}^i_A$ and obtain $\tilde{P}^i_A = \epsilon^2 11 m^2_A v^i_A/(3\epsilon R_A) + O(\epsilon^2)$ from the definition of P^i_A. Thus from the fact that $Q^i_A = \epsilon^2 m^2_A v^i_A/(6\epsilon R_A) + O(\epsilon^2)$ we find that the 0.5PN momentum–velocity relation does not depend on R_A and $\bar{P}^i_A = \bar{P}^\tau_A v^i_A + O(\epsilon^2)$. Finally evaluating the surface integrals in the general form of the equations of motion, using the "renormalized" (or barred) moments, we find that the equations of motion are independent of R_A, as expected. Thus from now on we shall use the same symbol for the renormalized moments as for the "bare" moments for notational simplicity.

3

The third post-Newtonian approximation

The final form of the 3PN equations is obtained. Technical points relating the free choice of dipole moments and the removal of logarithmic terms are examined. Special properties of the equations are listed and some comparisons with, and observations of, alternative approaches are given.

3.1 Super-potential versus surface integrals

We have so far been able to solve all of the Poisson equations necessary to derive the 2.5PN gravitational field. At 3PN order we encounter a situation in which a part of a solution of Poisson's equation constitutes what we call a super-potential. We take a super-potential to be a particular solution of a Poisson equation whose source term is non-compact. An example is given by

$$\frac{r_1^i r_1^j r_1^k r_2^l}{r_1^5 r_2^3} = \Delta\left[-\frac{1}{3}\frac{\partial^4}{\partial z_1^i\,\partial z_1^j\,\partial z_1^k\,\partial z_2^l} f^{(1,-1)} + \frac{1}{3}\left(\delta^{ij}\partial_{z_1^k} + \delta^{ik}\partial_{z_1^j} + \delta^{jk}\partial_{z_1^i}\right)\partial_{z_2^l}\ln S\right], \tag{3.1}$$

where $S = r_1 + r_2 + r_{12}$ and $f^{(1,-1)}$, which satisfies $\Delta f^{(1,-1)} = r_1/r_2$, is given in Jaranowski and Schäfer (1998*a*, 2001) as

$$f^{(1,-1)} = \frac{1}{18}\left(-r_1^2 - 3r_1 r_{12} - r_{12}^2 + 3r_1 r_2 + 3r_{12} r_2 + r_2^2\right) + \frac{1}{6}\left(-r_1^2 + r_{12}^2 + r_2^2\right)\ln S. \tag{3.2}$$

It is possible to add a homogeneous solution to super-potentials. In the case above we could add $1/r_1$ (say) to $f^{(1,-1)}$. In our formalism the only place where we have made use of super-potentials is in equation (2.75). It is easy to see that the contribution from an additional homogeneous solution cancels out between the "$-4\pi g(\vec{x})$" term and the surface integral in this equation. To evaluate surface integrals appearing in the general form of the equations of motion (2.109) requires super-potentials in the spatial region N/B which do not possess a singularity due to the point particle limit. Useful super-potentials for the derivation of the 3PN field are given in Jaranowski and Schäfer (1998*a*, 2001), Blanchet *et al.* (1998), Blanchet and Faye (2001*b*) and

Itoh (2004). A method that achieves this has been developed in Itoh (2004) for a source term of the form

$$\partial_{z_A^i}\partial_{z_{A'}^j} g(\vec{x}) \equiv \partial_{z_A^i}\partial_{z_{A'}^j}\left(\frac{(\ln r_1)^p(\ln r_2)^q}{r_1^a r_2^b}\right), \tag{3.3}$$

where a and b are integers and $p = 0, 1$, $q = 0, 1$. Note that $A, A' = 1, 2$. Then we take spatial derivatives out of the Poisson integral

$$\int_{N/B} \frac{d^3y}{|\vec{x}-\vec{y}|}\partial_{z_A^i}\partial_{z_{A'}^j} g(\vec{y}) = \partial_{z_A^i}\partial_{z_{A'}^j}\int_{N/B} d^3y\,\frac{g(\vec{y})}{|\vec{x}-\vec{y}|} + \partial_{z_A^i}\oint_{\partial B_{A'}} dS_j\,\frac{g(\vec{y})}{|\vec{x}-\vec{y}|}$$
$$+\oint_{\partial B_A} dS_i\,\frac{\partial_{z_{A'}^j} g(\vec{y})}{|\vec{x}-\vec{y}|}. \tag{3.4}$$

We note that the integration region is N/B and therefore $g(\vec{x})$ is nonsingular in N/B. For this class of source term a method is described in Itoh (2004) to find a field $F_{[A,c]}^{(m,n)}$ in the neighborhood of star A in the following sense:

$$\Delta F_{[A,c]}^{(p,m,q,n)} - (\ln r_1)^p r_1^m (\ln r_2)^q r_2^n = O\left(\frac{r_A^{c+1}}{r_{12}^{c+1}}\right), \qquad \text{as } r_A \to 0. \tag{3.5}$$

We have checked that at 3PN order the field resulting from this method is the same as the field obtained from the usual (super-potential) method whenever the super-potentials are available. However this method is not always adequate and so we turn to a different approach to derive the equations of motion.

In practice we are required to evaluate a surface integral of the Landau–Lifshitz pseudotensor which has an integrand of the form $h^{\mu\nu}{}_{,\alpha}h^{\lambda\sigma}{}_{,\beta}$. Thus roughly speaking we must evaluate a surface integral of the form (Newtonian potential) × (3PN potential) to find the 3PN equations of motion when it appears to be impossible to find the factor (3PN potential) in closed form. The surface integral has the form

$$\oint_{\partial B_1} dS_k \frac{r_A^l}{r_A^3} \operatorname*{disc}_{\epsilon R_A} \int_{N/B} d^3y\,\frac{f(\vec{y})}{|\vec{x}-\vec{y}|}, \tag{3.6}$$

for star 1. Here the operator $\mathrm{disc}_{\epsilon R_A}$ means "discard all the ϵR_A dependent terms other than logarithms of ϵR_A" (Itoh, 2004). The strategy to evaluate this type of integral is to exchange the order of integration so that we first calculate the surface integral and then calculate the volume integral. In the process of doing this we must put an operator $\mathrm{disc}_{\epsilon R_A}$ in front of the Poisson integral as in (3.6). Interchanging the order of integration results in

$$\oint_{\partial B_1} dS_k \frac{r_A^l}{r_A^3} \operatorname*{disc}_{\epsilon R_A} \int_{N/B} d^3y\,\frac{f(\vec{y}_1)}{|\vec{r}_1-\vec{y}_1|}$$
$$= \lim_{r_1'\to\epsilon R_1} \operatorname*{disc}_{\epsilon R_A} \int_{N/B} d^3y_1\, f(\vec{y}_1) \oint_{\partial B_1'} dS_k\,\frac{1}{|\vec{y}_1 - r_1'\vec{n}_1|}\partial_{z_A^l}\frac{1}{r_A}, \tag{3.7}$$

where we have defined a sphere B'_1, whose center is $\vec{z}_1$ and radius is r'_1, with r'_1 a constant which is slightly larger than ϵR_1 for any (small) ϵ, i.e. $\epsilon R_1 < r'_1 \ll r_{12}$.

The reason we have introduced r'_1 is as follows. Suppose that we want to consider an integrand for which the super-potential is available. By calculating the Poisson integral we then have a part of the field corresponding to the integrand. This part generally depends on ϵR_A. However we can discard such R_A-dependent terms (other than logarithmic dependence) as explained in Section 2.13. Using the resulting R_A-independent field we then evaluate the surface integrals in the general form of the 3PN equations of motion by discarding the ϵR_A dependence emerging from the surface integrals and obtain the equations of motion. The "discarding-ϵR_A" procedure must be employed every time the field is derived and *also* when deriving the equations of motion. The variable r'_1 was introduced to distinguish the two kinds of ϵR_A dependence and to discard ϵR_A dependence in the correct order. We give here a simple example. Consider the following integral:

$$\oint_{\partial B_1} dS_k \frac{r_1^k}{r_1^3} \int_{N/B} \frac{d^3y}{|\vec{x}-\vec{y}|} \frac{1}{y_1^2}. \tag{3.8}$$

Using $\Delta \ln r_1 = 1/r_1^2$ we can integrate the Poisson integral and obtain the "field"

$$\int_{N/B} \frac{d^3y}{|\vec{x}-\vec{y}|} \frac{1}{y_1^2} = -4\pi \ln\left(\frac{r_1}{\mathcal{R}/\epsilon}\right) + 4\pi - 4\pi\frac{\epsilon R_1}{r_1} + O\left((\epsilon R_A)^2\right). \tag{3.9}$$

Since the "body zone contribution" must have an ϵR_1 dependence hidden in the "moments" in the form $4\pi\epsilon R_1/r_1[+O((\epsilon R_A)^2)]$ (see Section 2.13), the last term should be discarded before we evaluate the surface integral in (3.8) using this "field". The surface integral then yields the "equations of motion"

$$16\pi^2\left(\ln\left(\frac{\mathcal{R}/\epsilon}{\epsilon R_1}\right) + 1\right). \tag{3.10}$$

On the other hand we can derive the "equations of motion" by first evaluating the surface integral over $\partial B'_1$ to obtain

$$\begin{aligned}\oint_{\partial B_1} dS_k \frac{r_1^k}{r_1^3} \int_{N/B} \frac{d^3y}{|\vec{x}-\vec{y}|} \frac{1}{y_1^2} &= \int_{N/B} \frac{d^3y}{y_1^2} \oint_{\partial B'_1} dS_k \frac{r_1^k}{r_1^3} \frac{1}{|\vec{r}_1 - \vec{y}_1|}, \\ &= 16\pi^2\left[\int_{\epsilon R_1}^{r'_1} \frac{dy}{r'_1} + \int_{r'_1}^{\mathcal{R}/\epsilon} \frac{dy}{y_1}\right], \\ &= 16\pi^2\left(\ln\left(\frac{\mathcal{R}/\epsilon}{r'_1}\right) + 1 - \frac{\epsilon R_1}{r'_1}\right). \end{aligned} \tag{3.11}$$

Thus if we take ∂B_1 as the integration region instead of $\partial B'_1$ in the first equality in (3.11), or if we take $\lim_{r'_1 \to \epsilon R_1}$ without employing $\mathrm{disc}_{\epsilon R_A}$ beforehand, we will obtain an incorrect result

$$16\pi^2 \ln\left(\frac{\mathcal{R}/\epsilon}{\epsilon R_1}\right), \tag{3.12}$$

because it disagrees with (3.10). With this caution in mind it is straightforward (though tedious) to evaluate the surface integrals and then the volume integrals, and thus evaluate all the necessary integrals, for the derivation of the 3PN equations of motion.

In deriving the 3PN equations of motion we have used the super-potential method whenever possible, and used a combination of the above three methods when necessary. In fact, for a computational check, we have used the direct-integration method to evaluate the contributions to the equations of motion from all of the 3PN nonretarded field components in N/B namely, ${}_8h^{\tau i}_{N/B\,n=0}$ and ${}_{10}h^{\tau\tau}_{N/B\,n=0} + {}_8h^{l}_{N/B\,n=0\,l}$. We obtain the same result from the two modes of computation; the result from the direct-integration method agrees with that from the combination of the three methods: the direct-integration method, the super-potential method, and the super-potential-in-series method.

3.2 3PN mass–energy relation

The direct-integration part does not play any role in the evaluation of the evolution equation of P^τ_A at 3PN order. Thus we can use the same method now as in the evaluation of the 2.5PN equations of motion. Evaluating the surface integrals in (2.108) we obtain the evolution equation for P^τ_A as

$$\begin{aligned}\frac{dP^\tau_{1\Theta}}{d\tau} &= -\epsilon^2\frac{m_1m_2}{r_{12}^2}\left[4(\vec{n}_{12}\cdot\vec{v}_1) - 3(\vec{n}_{12}\cdot\vec{v}_2)\right] + \epsilon^4\frac{m_1m_2}{r_{12}^2}f_1 \\ &\quad + \epsilon^6\frac{m_1m_2}{r_{12}}(f_2+f_3+f_4) + O(\epsilon^7). \end{aligned} \tag{3.13}$$

with

$$\begin{aligned} f_1 &= -\frac{9}{2}(\vec{n}_{12}\cdot\vec{v}_2)^3 + \frac{1}{2}v_1^2(\vec{n}_{12}\cdot\vec{v}_2) + 6(\vec{n}_{12}\cdot\vec{v}_1)(\vec{n}_{12}\cdot\vec{v}_2)^2 - 2v_1^2(\vec{n}_{12}\cdot\vec{v}_1) \\ &\quad +4(\vec{v}_1\cdot\vec{v}_2)(\vec{n}_{12}\cdot\vec{V}) + 5v_2^2(\vec{n}_{12}\cdot\vec{v}_2) - 4v_2^2(\vec{n}_{12}\cdot\vec{v}_1) \\ &\quad +\frac{m_1}{r_{12}}(-4(\vec{n}_{12}\cdot\vec{v}_2) + 6(\vec{n}_{12}\cdot\vec{v}_1)) + \frac{m_2}{r_{12}}(-10(\vec{n}_{12}\cdot\vec{v}_1) + 11(\vec{n}_{12}\cdot\vec{v}_2)), \end{aligned} \tag{3.14}$$

$$\begin{aligned} f_2 &= -\left(\frac{3}{2}v_1^4 + 2v_1^2v_2^2 + 4v_2^4\right)(\vec{n}_{12}\cdot\vec{v}_1) + \left(\frac{5}{8}v_1^4 + \frac{3}{2}v_1^2v_2^2 + 7v_2^4\right)(\vec{n}_{12}\cdot\vec{v}_2) \\ &\quad + \left(2v_1^2 + 4v_2^2\right)(\vec{n}_{12}\cdot\vec{v}_1)(\vec{v}_1\cdot\vec{v}_2) - \left(2v_1^2 + 8v_2^2\right)(\vec{n}_{12}\cdot\vec{v}_2)(\vec{v}_1\cdot\vec{v}_2) \\ &\quad + \left(3v_1^2 + 12v_2^2\right)(\vec{n}_{12}\cdot\vec{v}_1)(\vec{n}_{12}\cdot\vec{v}_2)^2 - \left(\frac{3}{4}v_1^2 + 12v_2^2\right)(\vec{n}_{12}\cdot\vec{v}_2)^3 \\ &\quad +2(\vec{n}_{12}\cdot\vec{v}_2)(\vec{v}_1\cdot\vec{v}_2)^2 - 6(\vec{n}_{12}\cdot\vec{v}_1)(\vec{n}_{12}\cdot\vec{v}_2)^2(\vec{v}_1\cdot\vec{v}_2) + 6(\vec{n}_{12}\cdot\vec{v}_2)^3(\vec{v}_1\cdot\vec{v}_2) \\ &\quad -\frac{15}{2}(\vec{n}_{12}\cdot\vec{v}_1)(\vec{n}_{12}\cdot\vec{v}_2)^4 - \frac{45}{8}(\vec{n}_{12}\cdot\vec{v}_2)^5, \end{aligned} \tag{3.15}$$

$$f_3 = \frac{m_1}{r_{12}}\Bigg\{\left(-42v_1^2 - \frac{117}{4}v_2^2\right)(\vec{n}_{12}\cdot\vec{v}_1) + 60(\vec{n}_{12}\cdot\vec{v}_1)^3 + \left(\frac{137}{4}v_1^2 + \frac{37}{2}v_2^2\right)(\vec{n}_{12}\cdot\vec{v}_2)$$
$$+\frac{297}{4}(\vec{n}_{12}\cdot\vec{v}_1)(\vec{v}_1\cdot\vec{v}_2) - \frac{219}{4}(\vec{n}_{12}\cdot\vec{v}_2)(\vec{v}_1\cdot\vec{v}_2) - 151(\vec{n}_{12}\cdot\vec{v}_1)^2(\vec{n}_{12}\cdot\vec{v}_2)$$
$$+109(\vec{n}_{12}\cdot\vec{v}_1)(\vec{n}_{12}\cdot\vec{v}_2)^2 - 23(\vec{n}_{12}\cdot\vec{v}_2)^3\Bigg\}, \tag{3.16}$$

and

$$f_4 = \frac{m_2}{r_{12}}\Bigg\{-\left(13v_1^2 + 18v_2^2\right)(\vec{n}_{12}\cdot\vec{v}_1) + \left(17v_1^2 + 25v_2^2\right)(\vec{n}_{12}\cdot\vec{v}_2)$$
$$+26(\vec{n}_{12}\cdot\vec{v}_1)(\vec{v}_1\cdot\vec{v}_2) - 28(\vec{n}_{12}\cdot\vec{v}_2)(\vec{v}_1\cdot\vec{v}_2) + 2(\vec{n}_{12}\cdot\vec{v}_1)^2(\vec{n}_{12}\cdot\vec{v}_2)$$
$$+16(\vec{n}_{12}\cdot\vec{v}_1)(\vec{n}_{12}\cdot\vec{v}_2)^2 - 20(\vec{n}_{12}\cdot\vec{v}_2)^3\Bigg\} + \frac{m_1^2}{r_{12}^2}\left(\frac{33}{4}(\vec{n}_{12}\cdot\vec{v}_1) - \frac{13}{2}(\vec{n}_{12}\cdot\vec{v}_2)\right)$$
$$-\frac{m_1 m_2}{r_{12}^2}\left(\frac{35}{4}(\vec{n}_{12}\cdot\vec{v}_1) + \frac{17}{4}(\vec{n}_{12}\cdot\vec{v}_2)\right) + \frac{m_2^2}{r_{12}^2}\left(-12(\vec{n}_{12}\cdot\vec{v}_1) + \frac{23}{2}(\vec{n}_{12}\cdot\vec{v}_2)\right). \tag{3.17}$$

Remarkably we can integrate (3.13) functionally to arrive at

$$P^{\tau}_{1\Theta} = m_1\left(1 + \epsilon^2\, {}_2\Gamma_1 + \epsilon^4\, {}_4\Gamma_1 + \epsilon^6\, {}_6\Gamma_1\right) + O(\epsilon^7), \tag{3.18}$$

with

$${}_2\Gamma_1 = \frac{1}{2}v_1^2 + \frac{3m_2}{r_{12}}, \tag{3.19}$$

$${}_4\Gamma_1 = -\frac{3m_2}{2r_{12}}(\vec{n}_{12}\cdot\vec{v}_2)^2 + \frac{2m_2}{r_{12}}v_2^2 + \frac{7m_2}{2r_{12}}v_1^2 - \frac{4m_2}{r_{12}}(\vec{v}_1\cdot\vec{v}_2) + \frac{3}{8}v_1^4$$
$$+\frac{7m_2^2}{2r_{12}^2} - \frac{5m_1m_2}{2r_{12}^2}, \tag{3.20}$$

and

$${}_6\Gamma_1 = \frac{m_1^2 m_2}{2r_{12}^3} + \frac{21 m_1 m_2^2}{4r_{12}^3} + \frac{5m_2^3}{2r_{12}^3} + \frac{5}{16}v_1^6 + \frac{m_2^2}{r_{12}^2}g_1 + \frac{m_1 m_2}{r_{12}^2}g_2 + \frac{m_2}{r_{12}}g_3, \tag{3.21}$$

with

$$g_1 = \frac{45}{4}v_1^2 + \frac{19}{2}v_2^2 + \frac{1}{2}(\vec{n}_{12}\cdot\vec{v}_1)^2 - 19(\vec{v}_1\cdot\vec{v}_2) - (\vec{n}_{12}\cdot\vec{v}_1)(\vec{n}_{12}\cdot\vec{v}_2)$$
$$-3(\vec{n}_{12}\cdot\vec{v}_2)^2, \tag{3.22}$$

$$g_2 = \frac{43}{8}v_1^2 + \frac{53}{8}v_2^2 - \frac{69}{8}(\vec{n}_{12}\cdot\vec{v}_1)^2 - \frac{53}{4}(\vec{v}_1\cdot\vec{v}_2)$$
$$+\frac{85}{4}(\vec{n}_{12}\cdot\vec{v}_1)(\vec{n}_{12}\cdot\vec{v}_2) - \frac{69}{8}(\vec{n}_{12}\cdot\vec{v}_2)^2, \tag{3.23}$$

$$g_3 = \frac{33}{8}v_1^4 + \frac{3}{2}v_1^2v_2^2 + v_2^4 - 6v_1^2(\vec{v}_1\cdot\vec{v}_2) - 4v_2^2(\vec{v}_1\cdot\vec{v}_2) - \frac{7}{4}v_1^2(\vec{n}_{12}\cdot\vec{v}_2)^2$$
$$-\frac{5}{2}v_2^2(\vec{n}_{12}\cdot\vec{v}_2)^2 + 2(\vec{v}_1\cdot\vec{v}_2)^2 + 2(\vec{n}_{12}\cdot\vec{v}_2)^2(\vec{v}_1\cdot\vec{v}_2)$$
$$+\frac{9}{8}(\vec{n}_{12}\cdot\vec{v}_2)^4. \tag{3.24}$$

Equation (3.18) together with (3.19), (3.20), and (3.21) constitutes the 3PN-order mass–energy relation.

3.3 The meaning of $P^{\tau}_{A\Theta}$

To elucidate the meaning of $P^{\tau}_{A\Theta}$ we begin by expanding in an ϵ series the 4-velocity of star A normalized by $g_{\mu\nu}u^{\mu}_Au^{\nu}_A = -\epsilon^{-2}$, where $u^i_A = u^{\tau}_Av^i_A$. We find that

$$u^{\tau}_A = 1 + \epsilon^2\left[\frac{1}{2}v_A^2 + \frac{1}{4}{}_4h^{\tau\tau}\right]$$
$$+\epsilon^4\left[\frac{1}{4}{}_6h^{\tau\tau} + \frac{1}{4}{}_4h^k{}_k - \frac{3}{32}({}_4h^{\tau\tau})^2 + \frac{5}{8}{}_4h^{\tau\tau}v_A^2 - {}_4h^{\tau}{}_kv^k_A + \frac{3}{8}v_A^4\right]$$
$$+\epsilon^5\frac{1}{4}\left[{}_7h^{\tau\tau} + {}_5h^k{}_k\right]$$
$$+\epsilon^6\left[\frac{1}{4}{}_8h^{\tau\tau} + \frac{1}{4}{}_6h^k{}_k + \frac{1}{16}{}_4h^{\tau\tau}{}_4h^k{}_k - \frac{3}{16}{}_4h^{\tau\tau}{}_6h^{\tau\tau}\right.$$
$$+\frac{7}{128}({}_4h^{\tau\tau})^3 + \frac{1}{4}{}_4h^{\tau k}{}_4h^{\tau}{}_k$$
$$-{}_6h^{\tau}{}_kv^k_A - \frac{1}{4}{}_4h^{\tau\tau}{}_4h^{\tau}{}_kv^k_A + \frac{1}{2}{}_4h_{kl}v^k_Av^l_A + \frac{1}{8}{}_4h^k{}_kv_A^2 + \frac{5}{64}({}_4h^{\tau\tau})^2v_A^2$$
$$\left.+\frac{5}{8}{}_6h^{\tau\tau}v_A^2 - \frac{3}{2}{}_4h^{\tau}{}_kv^k_Av_A^2 + \frac{27}{32}{}_4h^{\tau\tau}v_A^4 + \frac{5}{16}v_A^6\right]$$
$$+O(\epsilon^7). \tag{3.25}$$

The field variable $h^{\mu\nu}$ appearing here should be evaluated at $\vec{z}_A$. However this is a formal series since the metric derived via the point particle description diverges at $\vec{z}_A$. Let us regularize this equation with the Hadamard partie finie regularization [see Hadamard (1932), Schwartz (1966) and, for example, Blanchet and Faye (2000*a*) and Blanchet *et al.* (1998) in the literature of the post-Newtonian approximation]. Consider a function $f(\vec{x})$ which can be expanded around $\vec{z}_1$ in the form

$$f(\vec{r}_1) = \sum_{p=p_0}\frac{1}{r_1^p}\hat{f}_p(\vec{N}_1). \tag{3.26}$$

Then the Hadamard partie finie at $\vec{z}_1$ of the function $f(\vec{x})$ is defined by

$$\oint \frac{d\Omega_{\mathbf{n_1}}}{4\pi} \underset{0}{\hat{f}}(\vec{n}_1). \tag{3.27}$$

For example in this procedure $h^{\tau\tau}$ becomes [see (2.128)]

$$[h^{\tau\tau}]_1^H = 4\epsilon^4 \frac{P_2^\tau}{r_{12}} + \epsilon^6 \left[-2\frac{m_2}{r_{12}} \left\{ (\vec{n}_{12} \cdot \vec{v}_2)^2 - v_2^2 \right\} - 16\frac{m_1 m_2}{r_{12}^2} + 7\frac{m_2^2}{r_{12}^2} \right] + O(\epsilon^7), \tag{3.28}$$

for star 1. In the above equation $[f]_A^{\rm H}$ means that we regularize the quantity f at star A by the Hadamard partie finie. Evaluating (3.25) and $[\sqrt{-g}]_A^{\rm H}$ by this procedure, then comparing the result to (3.18) combined with (3.19), (3.20), and (3.21), we find that up to 3PN order,

$$P_{A\Theta}^\tau = m_A \left[\sqrt{-g} u_A^\tau \right]_A^{\rm H}. \tag{3.29}$$

It is important to note that even after regularizing all the divergent terms there remains a nonlinear effect. Equation (3.29) is what one would expect although it has not been necessary to assume that such a relation exists. It has been derived by solving the evolution equation for $P_{A\Theta}^\tau$ functionally. The fact that (3.29) has an expected form supports the use of the Hadamard partie finie regularization (or any other regularization procedure if it can reproduce (3.29)) to derive the 3PN mass–energy relation to deal with divergences when one uses Dirac delta distributions. In practice we need the 2.5PN field and a part of the 3PN field (${}_8h^{\tau[i,j]}$) to evaluate the 3PN mass–energy relation. This does not mean that the Hadamard partie finie regularization is sufficient to derive the 3PN equations of motion. Indeed when Dirac delta functions are used to describe the point particle limit instead of the strong field point particle limit, one needs another regularization scheme, such as the dimensional regularization scheme, to derive the 3PN equations of motion in an unambiguous form (Blanchet *et al.*, 2004). We note that up to 3PN order

$$\left[\sqrt{-g} u_A^\tau \right]_A^{\rm H} = [\sqrt{-g}]_A^{\rm H} [u_A^\tau]_A^{\rm H} + O(\epsilon^7), \tag{3.30}$$

is satisfied if we use the Hadamard partie finie regularization explained above.

3.4 3PN momentum–velocity relation

We derive the 3PN momentum–velocity relation by calculating the Q_A^i integral at 3PN order. From the definition of the Q_A^i integral (2.89)

$$Q_A^i = \epsilon^6 \oint_{\partial B_A} dS_k \left({}_{10}\Lambda_N^{\tau k} - v_A^k {}_{10}\Lambda_N^{\tau\tau} \right) y_A^i, \tag{3.31}$$

we see that the calculation required is almost the same as that for the evaluation of $dP_A^\tau/d\tau$. We do not need to use the direct-integration method to compute the Q_A^i integral. Therefore it is straightforward to evaluate the surface integrals in the definition of Q_A^i. We split Q_A^i into $Q_{A\Theta}^i$ and $Q_{A\chi}^i$ as

$$Q_A^i = Q_{A\Theta}^i + Q_{A\chi}^i, \tag{3.32}$$

with

$$Q^i_{A\Theta} = \epsilon^{-4} \oint_{\partial B_A} dS_k \left(\Theta_N^{\tau k} - v_A^k \Theta_N^{\tau\tau}\right) y_A^i, \tag{3.33}$$

$$Q^i_{A\chi} = \epsilon^{-4} \oint_{\partial B_A} dS_k \left(\chi_N^{\tau k\alpha\beta}{}_{,\alpha\beta} - v_A^k \chi_N^{\tau\tau\alpha\beta}{}_{,\alpha\beta}\right) y_A^i. \tag{3.34}$$

We have explicitly $Q^i_{A\Theta}$ given by

$$_{\le 6}Q^i_{1\Theta} = -\epsilon^6 \frac{m_1^3 m_2 n_{12}^{\langle ij\rangle} v_{12}^j}{2r_{12}^3} = -\epsilon^6 \frac{d}{d\tau}\left(\frac{m_1^3 m_2}{6r_{12}^3} r_{12}^i\right) = \epsilon^6 \frac{d}{d\tau}\left(\frac{1}{6} m_1^3 a_1^i\right), \tag{3.35}$$

where $_{\le n}f$ stands for the part of f up to order n inclusive. We note that a_A^i in the last expression is evaluated with the Newtonian acceleration. $Q_A^i = O(\epsilon^6)$ appears at the 4PN or higher-order field. Thus up to 3PN order $_6Q_A^i$ enters the equations of motion only through the 3PN momentum–velocity relation. For this reason $Q^i_{A\Theta}$, rather than $Q^i_{A\chi}$, is necessary to derive the 3PN equations of motion. The explicit expression for $Q^i_{A\chi}$ is given in Itoh (2004).

Now with $Q^i_{A\Theta}$ in hand we set out to derive the momentum–velocity relation. It turns out that the χ part of the momentum–velocity relation is an identity. Defining the Θ parts of P_A^μ and D_A^i in the same way as for Q_A^i, it turns out that $P^i_{A\chi} = P^\tau_{A\chi} v_A^i + Q^i_{A\chi} + \epsilon^2 d^2 D^i_{A\chi}/d\tau^2$ is an identity (Itoh, 2004). Note that $P^\mu_{A\chi}$ and $D^i_{A\chi}$ can be converted into surface integrals and hence can be evaluated explicitly. Defining the Θ parts of P_A^μ and D_A^i in the same way as for Q_A^i we obtain

$$P^i_{1\Theta} = P^\tau_{1\Theta} v_1^i + \epsilon^6 \frac{d}{d\tau}\left(\frac{1}{6} m_1^3 a_1^i\right) + \epsilon^2 \frac{dD^i_{1\Theta}}{d\tau}. \tag{3.36}$$

As explained in Chapter 2, we define the representative point z_A^i of star A by choosing the value of D_A^i. It is natural to expect that $\vec{z}_A$ is in star A unless the star is, say, crescent shaped. By the same token it would not be useful to choose the representative point of the sun to be in the earth. One may set D_A^i equal to zero up to 2.5PN order. Alternatively one may find it "natural" to want a 3-momentum proportional to a 3-velocity and make the choice of

$$D^i_{A\Theta} = -\epsilon^4 \frac{1}{6} m_A^3 a_A^i = \epsilon^4 \delta^i_{A\Theta}. \tag{3.37}$$

Henceforth we shall define z_A^i by this equation.

It is important to realize that the nonzero dipole moment D_A^i of order ϵ^4 affects the 3PN field and the 3PN equations of motion in essentially the same way as the Newtonian dipole moment affects the Newtonian field and equations of motion. From equations (10.4)–(10.6) of Chapter 2 we see that $\delta^i_{A\Theta}$ appears only at $_{10}h^{\tau\tau}$ as

$$h^{\tau\tau}|_{\delta_{A\Theta}} = 4\epsilon^{10} \sum_{A=1,2} \frac{\delta^k_{A\Theta} r_A^k}{r_A^3} + O(\epsilon^{11}). \tag{3.38}$$

The corresponding acceleration becomes

$$m_1 a_1^i|_{\delta_{A\Theta}} = -\epsilon^6 \frac{3m_1 \delta_{2\Theta}^i}{r_{12}^3} n_{12}^{\langle ik \rangle} + \epsilon^6 \frac{3m_2 \delta_{1\Theta}^i}{r_{12}^3} n_{12}^{\langle ik \rangle} - \epsilon^6 \frac{d^2 \delta_{1\Theta}^i}{d\tau^2}. \tag{3.39}$$

The last term compensates for the Q_A^i integral contribution appearing through the momentum–velocity relation (3.36). Note that this change of the acceleration does not affect the existence of the conservation of the (Newtonian) energy,

$$m_1 a_1^i \,|_{\delta_{A\Theta}} v_1^i + m_2 a_2^i \big|_{\delta_{A\Theta}} v_2^i = \epsilon^6 \frac{d}{d\tau} \sum_{A=1,2} \left[\delta_{A\Theta}^k \frac{dv_A^k}{d\tau} - v_A^k \frac{d}{d\tau} \delta_{A\Theta}^k \right]. \tag{3.40}$$

3.5 3PN equations of motion with log terms

To derive the 3PN equations of motion we must evaluate the surface integrals in the general form of the equations of motion given in (2.109) using the field ${}_8h^{\tau\tau}$, the field ${}_{\le 6}h^{\mu\nu}$, the 3PN body zone contributions, and the 3PN in the region N/B contributions corresponding to the results from the super-potential method and the super-potential-in-series method. We then combine the result with the terms from the direct-integration method. From 3PN order the $Q_A^{K_l i}$ and $R_A^{K_l ij}$ integrals appearing in the 3PN field in $h_B^{\mu\nu}$ contribute to the 3PN equations of motion. ${}_6Q_{A\Theta}^i$ given in (3.35) affects the 3PN equations of motion through the 3PN momentum–velocity relation. Since we define the representative points of the stars via (3.37) we must add the corresponding acceleration given by (3.39). Furthermore our choice of the representative points of the stars makes $D_{A\chi}^i$ appear independently of $D_{A\Theta}^i$ in the field, and hence ${}_4D_{A\chi}^i$ influences the 3PN equations of motion. In summary the ${}_{\le 4}Q_A^{K_l i}$, ${}_{\le 4}R_A^{K_l ij}$, ${}_6Q_{A\Theta}^i$, $\delta_{A\Theta}^i$, and ${}_4D_{A\chi}^i$ contributions to the 3PN field can be written as

$${}_{10}h^{\tau\tau} + {}_8h^k{}_k = 4 \sum_{A=1,2} \frac{r_A^k}{r_A^3} \left(\delta_{A\Theta}^k + {}_4D_{A\chi}^k + {}_4R_A^{kll} - \frac{1}{2} {}_4R_A^{llk} \right) + \dots, \tag{3.41}$$

where the dots stand for other contributions. On the other hand ${}_6Q_{A\Theta}^i$ and $\delta_{A\Theta}^i$ contribute to the equations of motion through the momentum–velocity relation in [compare (2.109)]

$$m_1 \left(\frac{dv_1^i}{d\tau} \right)_{\le 3\text{PN}} = -\epsilon^6 \frac{d_6 Q_{1\Theta}^i}{d\tau} - \epsilon^6 \frac{d^2 \delta_{1\Theta}^i}{d\tau^2} + \dots, \tag{3.42}$$

but they cancel each other out on account of the choice (3.37). Then these contributions to the 3PN acceleration can be summarized by

$$\begin{pmatrix} \text{the contribution to } m_1 a_1^i \text{ from} \\ {}_{\le 4}Q_A^{K_l i}, {}_{\le 4}R_A^{K_l ij}, {}_6Q_{A\Theta}^i, \delta_{A\Theta}^i, {}_4D_{A\chi}^i \end{pmatrix} = \epsilon^6 \frac{118}{9} \frac{m_1^3 m_2^2}{r_{12}^6} r_{12}^i + \epsilon^6 \frac{118}{9} \frac{m_1^2 m_2^3}{r_{12}^6} r_{12}^i. \tag{3.43}$$

Collecting all of the contributions mentioned above we obtain the 3PN equations of motion. However we find that logarithmic terms having the arbitrary constants

ϵR_A in their arguments survive. These logarithm-dependent terms are given explicitly by

$$
\begin{aligned}
& m_1 \left(\frac{dv_1^i}{d\tau} \right)^{\text{with log}} \\
&= m_1 \left(\frac{dv_1^i}{d\tau} \right)_{\leq 2.5\text{PN}} \\
&+ \epsilon^6 \frac{m_1^2 m_2}{r_{12}^3} \left[\frac{44 m_1^2}{3 r_{12}^2} n_{12}^i \ln\left(\frac{r_{12}}{\epsilon R_1} \right) - \frac{44 m_2^2}{3 r_{12}^2} n_{12}^i \ln\left(\frac{r_{12}}{\epsilon R_2} \right) \right. \\
&\qquad \left. + \frac{22 m_1}{r_{12}} \left(5 (\vec{n}_{12} \cdot \vec{V})^2 n_{12}^i - V^2 n_{12}^i - 2 (\vec{n}_{12} \cdot \vec{V}) V^i \right) \ln\left(\frac{r_{12}}{\epsilon R_1} \right) \right] \\
&+ \cdots + O(\epsilon^7),
\end{aligned}
\tag{3.44}
$$

where the acceleration through 2.5PN order, $\left(dv_1^i/d\tau\right)_{\leq 2.5\text{PN}}$, is the Damour and Deruelle (1981*a*, 1981*b*, 1981*c*) 2.5PN acceleration. It has also been computed in (Itoh *et al.*, 2001). The dots stand for the terms that do not include any logarithms. Since this equation contains two arbitrary constants, the body zone radii R_A, at first sight its predictive power for the orbital motion of the binary seems to be limited. However we shall demonstrate below that by a reasonable redefinition of the representative points of the stars we can remove R_A from the equations of motion.

3.6 The arbitrary constant ϵR_A

The logarithms appearing in the 3PN equations of motion (3.44) are in a sense easy to understand. Since the post-Newtonian approximation is a weak field expansion, at some level of iteration in the evaluation of the field, logarithms will inevitably appear in the field, originating in volume integrals such as

$$
\int_{N/B} \frac{d^3 y}{|\vec{x} - \vec{y}|} \left(\frac{m}{y} \right)^n,
\tag{3.45}
$$

where m and y are the typical mass and length scales respectively in the orbital motion and n is a positive integer. For example with $m_1^3 (\vec{r}_1 \cdot \vec{a}_1)/r_1^5$ as an integrand we find

$$
\int_{N/B} \frac{d^3 y}{|\vec{x} - \vec{y}|} \frac{m_1^3 (\vec{y}_1 \cdot \vec{a}_1)}{y_1^5} = \frac{m_1^3 (\vec{r}_1 \cdot \vec{a}_1)}{3 r_1^3} \ln\left(\frac{r_1}{\epsilon R_1} \right) + \dots .
\tag{3.46}
$$

We could remove the ϵR_A dependence in the 3PN equations of motion by an alternative choice of the center of mass. For example the following choice of the representative point of star A removes the ϵR_A dependence in (3.44):

$$
D^i_{A\Theta,\text{new}}(\tau) = \epsilon^4 \delta^i_{A\Theta}(\tau) - \epsilon^4 \frac{22}{3} m_A^3 a_A^i \ln\left(\frac{r_{12}}{\epsilon R_A} \right) \equiv \epsilon^4 \delta^i_{A\Theta}(\tau) + \epsilon^4 \delta^i_{A\,\text{ln}}(\tau) \equiv \epsilon^4 \delta^i_A(\tau).
\tag{3.47}
$$

This redefinition of the center of mass does not affect the energy conservation (3.40). We can examine the effect of this redefinition on the equations of motion using (3.39) (using $\delta^i_{A\,\ln}$ instead of $\delta_{A\Theta}$). Then we have

$$\begin{aligned} m_1 a_1^i|_{\delta_{A\,\ln}} &= -\epsilon^6 \frac{3m_1\delta^i_{2\,\ln}}{r_{12}^3} n_{12}^{\langle ik\rangle} + \epsilon^6 \frac{3m_2\delta^i_{1\,\ln}}{r_{12}^3} n_{12}^{\langle ik\rangle} - \epsilon^6 \frac{d^2\delta^i_{1\,\ln}}{d\tau^2}, \\ &= -\frac{44m_1^3m_2^2}{3r_{12}^5} n_{12}^i \ln\left(\frac{r_{12}}{\epsilon R_1}\right) + \frac{44m_1^2m_2^3}{3r_{12}^5} n_{12}^i \ln\left(\frac{r_{12}}{\epsilon R_2}\right) \\ &\quad -\frac{22m_1^3m_2}{r_{12}^4}\left(5(\vec{n}_{12}\cdot\vec{V})^2 n_{12}^i - V^2 n_{12}^i - 2(\vec{n}_{12}\cdot\vec{V})V^i\right)\ln\left(\frac{r_{12}}{\epsilon R_1}\right) \\ &\quad +\frac{22m_1^3m_2}{3r_{12}^4}\left(\frac{m_1}{r_{12}}n_{12}^i + \frac{m_2}{r_{12}}n_{12}^i - V^2 n_{12}^i + 8(\vec{n}_{12}\cdot\vec{V})^2 n_{12}^i \right. \\ &\qquad \left. -2(\vec{n}_{12}\cdot\vec{V})^2 V^i\right). \end{aligned} \tag{3.48}$$

Comparing this with (3.44) we easily see that if the representative point z_A^i of star A is defined by

$$D^i_{A\Theta,\mathrm{new}}(\tau) = \epsilon^{-6}\int_{B_A} d^3y\ (y^i - z_A^i(\tau))\ \Theta_N^{\tau\tau}(\tau, y^k) = \epsilon^4\delta_A^i(\tau), \tag{3.49}$$

then z_A^i satisfies equations of motion which are free from logarithmic terms and thus free of ambiguity up to 3PN order inclusive. We emphasize that in our formalism z_A^i is defined by the value of D_A^i, and in turn we have a freedom to assign to D_A^i any value we like, although it is natural to choose a value of D_A^i such that z_A^i lies inside star A. We also note that z_A^i is defined order by order.

We remark here that Blanchet and Faye (2001*b*) have already noticed that in their 3PN equations of motion a suitable coordinate transformation removes parts of the logarithmic dependence on arbitrary parameters corresponding roughly to our body zone radii. However unlike the case discussed here their coordinate transformation does not remove the logarithmic dependence of their free parameters completely. The remaining logarithmic dependence was used by them in order to make their equations of motion conservative. It is well known that choosing different values of the dipole moments corresponds to coordinate transformations.

Our choice of the dipole moments of the stars is in a sense natural. To see this consider the harmonic condition

$$\begin{aligned} 0 = h^{\tau\mu}{}_{,\mu} &= 4\epsilon^4 \sum_{A=1,2}\left[\frac{\dot{P}_A^\tau}{r_A} + \frac{r_A^i}{r_A^3}\left(P_A^\tau v_A^i + \epsilon^2 \dot{D}_A^i - P_A^i\right)\right] \\ &\quad + \sum_{A=1,2}\oint_{\partial B_A}\frac{dS_i}{|\vec{x}-\vec{y}|}\left(\Lambda_N^{\tau i} - v_A^i\Lambda_N^{\tau\tau}\right) + \dots, \end{aligned} \tag{3.50}$$

$$0 = h^{i\mu}{}_{,\mu} = 4\epsilon^4\sum_{A=1,2}\frac{\dot{P}_A^i}{r_A} + \sum_{A=1,2}\oint_{\partial B_A}\frac{dS_j}{|\vec{x}-\vec{y}|}\left(\Lambda_N^{ij} - v_A^j\Lambda_N^{\tau i}\right) + \dots, \tag{3.51}$$

where the dots following the plus signs represent irrelevant terms. These equations are a manifestation of the fact that the harmonic condition is consistent with the evolution equation for P_A^τ, the momentum–velocity relation, and the equations of motion (and relations among higher multipole moments hidden in the dots). Thus if the logarithmic dependence of ϵR_A arises from the second term of (3.51) then P_A^i must have the same logarithmic dependence (times a minus sign) to ensure harmonicity. This and the momentum–velocity relation in turn mean that P_A^τ, $v_A^i = \dot{z}_A^i$ or D_A^i have corresponding logarithmic dependence. Since we already know that the P_A^τ have no logarithmic dependence up to 3PN order, z_A^i or D_A^i should have logarithmic terms. This is consistent with the fact that a choice of D_A^i determines z_A^i and, as we have seen, z_A^i depends on logarithms if the old choice is made, while it does not if the new choice is made.

There is yet another fact which supports our interpretation. Let us retain $D_A^i \neq 0$ for the moment. We then find that the near zone dipole moment D_N^i, defined by a volume integral of $\Lambda_N^{\tau\tau} y^i$, becomes

$$\epsilon^2 D_N^i \equiv \epsilon^{-4} \int_N d^3y\, \Lambda_N^{\tau\tau} y^i = \sum_{A=1,2} P_A^\tau z_A^i + \epsilon^2 \sum_{A=1,2} D_A^i + \epsilon^{-4} \int_{N/B} d^3y\, \Lambda_N^{\tau\tau} y^i. \quad (3.52)$$

If we make the old choice of D_A^i then the volume integral here is given by

$$\int_{N/B} d^3y\, \Lambda_N^{\tau\tau} y^i = \epsilon^4 \frac{22}{3} \sum_{A=1,2} m_A^3 a_A^i \ln\left(\frac{r_{12}}{\epsilon R_A}\right) + \ldots, \quad (3.53)$$

where terms denoted by the dots have no logarithmic dependence. Note that the near zone dipole moment can be freely determined, as $D_N^i = 0$ (say), because we can define the origin of the near zone freely. By taking temporal derivatives twice of D_N^i, we see that $D_{A,\mathrm{new}}^i$ gives a natural definition of the center of the mass in terms of which the 3PN equations of motion are independent of ϵR_A.

3.7 Final form of 3PN equations of motion

By adding $m_1 a_1^i|_{\delta_A \ln}$ to (3.44) we obtain 3PN equations of motion, in the harmonic gauge, for two spherical compact stars whose representative points are defined by (3.49):

$$\begin{aligned} m_1 \frac{dv_1^i}{d\tau} &= -\frac{m_1 m_2}{r_{12}^2} n_{12}^i \\ &+ \epsilon^2 \frac{m_1 m_2}{r_{12}^2} n_{12}^i \left[-v_1^2 - 2v_2^2 + 4(\vec{v}_1 \cdot \vec{v}_2) + \frac{3}{2}(\vec{n}_{12} \cdot \vec{v}_2)^2 + \frac{5m_1}{r_{12}} + \frac{4m_2}{r_{12}} \right] \\ &+ \epsilon^2 \frac{m_1 m_2}{r_{12}^2} V^i \left[4(\vec{n}_{12} \cdot \vec{v}_1) - 3(\vec{n}_{12} \cdot \vec{v}_2) \right] + \epsilon^4 \frac{m_1 m_2}{r_{12}^2} \left(n_{12}^i \mathcal{T}_1 + V^i \mathcal{T}_2 \right) \\ &+ \epsilon^5 \frac{4m_1^2 m_2}{5 r_{12}^3} \left[n_{12}^i (\vec{n}_{12} \cdot \vec{V}) \left(-6\frac{m_1}{r_{12}} + \frac{52}{3}\frac{m_2}{r_{12}} + 3V^2 \right) \right. \\ &\left. + V^i \left(2\frac{m_1}{r_{12}} - 8\frac{m_2}{r_{12}} - V^2 \right) \right] + \epsilon^6 \frac{m_1 m_2}{r_{12}^2} \left(n_{12}^i \mathcal{T}_3 + V^i \mathcal{T}_4 \right) + O(\epsilon^7), \end{aligned} \quad (3.54)$$

where $\mathcal{T}_1$, $\mathcal{T}_2$, $\mathcal{T}_3$, $\mathcal{T}_4$ are given by

$$\begin{aligned}\mathcal{T}_1 = &-2v_2^4 + 4v_2^2(\vec{v}_1\cdot\vec{v}_2) - 2(\vec{v}_1\cdot\vec{v}_2)^2 + \frac{3}{2}v_1^2(\vec{n}_{12}\cdot\vec{v}_2)^2 + \frac{9}{2}v_2^2(\vec{n}_{12}\cdot\vec{v}_2)^2\\
&-6(\vec{v}_1\cdot\vec{v}_2)(\vec{n}_{12}\cdot\vec{v}_2)^2 - \frac{15}{8}(\vec{n}_{12}\cdot\vec{v}_2)^4 - \frac{57}{4}\frac{m_1^2}{r_{12}^2} - 9\frac{m_2^2}{r_{12}^2} - \frac{69}{2}\frac{m_1 m_2}{r_{12}^2}\\
&+\frac{m_1}{r_{12}}\Bigg(-\frac{15}{4}v_1^2 + \frac{5}{4}v_2^2 - \frac{5}{2}(\vec{v}_1\cdot\vec{v}_2) + \frac{39}{2}(\vec{n}_{12}\cdot\vec{v}_1)^2\\
&-39(\vec{n}_{12}\cdot\vec{v}_1)(\vec{n}_{12}\cdot\vec{v}_2) + \frac{17}{2}(\vec{n}_{12}\cdot\vec{v}_2)^2\Bigg)\\
&+\frac{m_2}{r_{12}}\left(4v_2^2 - 8(\vec{v}_1\cdot\vec{v}_2) + 2(\vec{n}_{12}\cdot\vec{v}_1)^2 - 4(\vec{n}_{12}\cdot\vec{v}_1)(\vec{n}_{12}\cdot\vec{v}_2) - 6(\vec{n}_{12}\cdot\vec{v}_2)^2\right),\end{aligned} \tag{3.55}$$

$$\begin{aligned}\mathcal{T}_2 = &\frac{m_1}{r_{12}}\left(-\frac{63}{4}(\vec{n}_{12}\cdot\vec{v}_1) + \frac{55}{4}(\vec{n}_{12}\cdot\vec{v}_2)\right) + \frac{m_2}{r_{12}}\left(-2(\vec{n}_{12}\cdot\vec{v}_1) - 2(\vec{n}_{12}\cdot\vec{v}_2)\right)\\
&+v_1^2(\vec{n}_{12}\cdot\vec{v}_2) + 4v_2^2(\vec{n}_{12}\cdot\vec{v}_1) - 5v_2^2(\vec{n}_{12}\cdot\vec{v}_2) - 4(\vec{v}_1\cdot\vec{v}_2)(\vec{n}_{12}\cdot\vec{V})\\
&-6(\vec{n}_{12}\cdot\vec{v}_1)(\vec{n}_{12}\cdot\vec{v}_2)^2 + \frac{9}{2}(\vec{n}_{12}\cdot\vec{v}_2)^3,\end{aligned} \tag{3.56}$$

$$\begin{aligned}\mathcal{T}_3 = &\frac{35}{16}(\vec{n}_{12}\cdot\vec{v}_2)^6 - \frac{15}{8}(\vec{n}_{12}\cdot\vec{v}_2)^4 v_1^2 + \frac{15}{2}(\vec{n}_{12}\cdot\vec{v}_2)^4(\vec{v}_1\cdot\vec{v}_2)\\
&+3(\vec{n}_{12}\cdot\vec{v}_2)^2(\vec{v}_1\cdot\vec{v}_2)^2 - \frac{15}{2}(\vec{n}_{12}\cdot\vec{v}_2)^4 v_2^2 + \frac{3}{2}(\vec{n}_{12}\cdot\vec{v}_2)^2 v_1^2 v_2^2\\
&-12(\vec{n}_{12}\cdot\vec{v}_2)^2(\vec{v}_1\cdot\vec{v}_2)v_2^2 - 2(\vec{v}_1\cdot\vec{v}_2)^2 v_2^2 + \frac{15}{2}(\vec{n}_{12}\cdot\vec{v}_2)^2 v_2^4\\
&+4(\vec{v}_1\cdot\vec{v}_2)v_2^4 - 2v_2^6\\
&+\frac{m_1}{r_{12}}\Bigg\{-\frac{171}{8}(\vec{n}_{12}\cdot\vec{v}_1)^4 + \frac{171}{2}(\vec{n}_{12}\cdot\vec{v}_1)^3(\vec{n}_{12}\cdot\vec{v}_2)\\
&-\frac{723}{4}(\vec{n}_{12}\cdot\vec{v}_1)^2(\vec{n}_{12}\cdot\vec{v}_2)^2 + \frac{383}{2}(\vec{n}_{12}\cdot\vec{v}_1)(\vec{n}_{12}\cdot\vec{v}_2)^3\\
&-\frac{455}{8}(\vec{n}_{12}\cdot\vec{v}_2)^4 + \frac{229}{4}(\vec{n}_{12}\cdot\vec{v}_1)^2 v_1^2 - \frac{205}{2}(\vec{n}_{12}\cdot\vec{v}_1)(\vec{n}_{12}\cdot\vec{v}_2)v_1^2\\
&+\frac{191}{4}(\vec{n}_{12}\cdot\vec{v}_2)^2 v_1^2 - \frac{91}{8}v_1^4 - \frac{229}{2}(\vec{n}_{12}\cdot\vec{v}_1)^2(\vec{v}_1\cdot\vec{v}_2)\\
&+244(\vec{n}_{12}\cdot\vec{v}_1)(\vec{n}_{12}\cdot\vec{v}_2)(\vec{v}_1\cdot\vec{v}_2) - \frac{225}{2}(\vec{n}_{12}\cdot\vec{v}_2)^2(\vec{v}_1\cdot\vec{v}_2)\\
&+\frac{91}{2}v_1^2(\vec{v}_1\cdot\vec{v}_2) - \frac{177}{4}(\vec{v}_1\cdot\vec{v}_2)^2 + \frac{229}{4}(\vec{n}_{12}\cdot\vec{v}_1)^2 v_2^2\end{aligned}$$

$$
\begin{aligned}
&- \frac{283}{2}(\vec{n}_{12}\cdot\vec{v}_1)(\vec{n}_{12}\cdot\vec{v}_2)v_2^2 + \frac{259}{4}(\vec{n}_{12}\cdot\vec{v}_2)^2 v_2^2 - \frac{91}{4}v_1^2 v_2^2 \\
&+ 43(\vec{v}_1\cdot\vec{v}_2)v_2^2 - \frac{81}{8}v_2^4\bigg\} \\
&+ \frac{m_2}{r_{12}}\{-6(\vec{n}_{12}\cdot\vec{v}_1)^2(\vec{n}_{12}\cdot\vec{v}_2)^2 + 12(\vec{n}_{12}\cdot\vec{v}_1)(\vec{n}_{12}\cdot\vec{v}_2)^3 \\
&+ 6(\vec{n}_{12}\cdot\vec{v}_2)^4 + 4(\vec{n}_{12}\cdot\vec{v}_1)(\vec{n}_{12}\cdot\vec{v}_2)(\vec{v}_1\cdot\vec{v}_2) \\
&+ 12(\vec{n}_{12}\cdot\vec{v}_2)^2(\vec{v}_1\cdot\vec{v}_2) + 4(\vec{v}_1\cdot\vec{v}_2)^2 \\
&- 4(\vec{n}_{12}\cdot\vec{v}_1)(\vec{n}_{12}\cdot\vec{v}_2)v_2^2 - 12(\vec{n}_{12}\cdot\vec{v}_2)^2 v_2^2 - 8(\vec{v}_1\cdot\vec{v}_2)v_2^2 + 4v_2^4\} \\
&+ \frac{m_2^2}{r_{12}^2}\bigg\{-(\vec{n}_{12}\cdot\vec{v}_1)^2 + 2(\vec{n}_{12}\cdot\vec{v}_1)(\vec{n}_{12}\cdot\vec{v}_2) + \frac{43}{2}(\vec{n}_{12}\cdot\vec{v}_2)^2 \\
&+ 18(\vec{v}_1\cdot\vec{v}_2) - 9v_2^2\bigg\} \\
&+ \frac{m_1 m_2}{r_{12}^2}\bigg\{\frac{415}{8}(\vec{n}_{12}\cdot\vec{v}_1)^2 - \frac{375}{4}(\vec{n}_{12}\cdot\vec{v}_1)(\vec{n}_{12}\cdot\vec{v}_2) + \frac{1113}{8}(\vec{n}_{12}\cdot\vec{v}_2)^2 \\
&- \frac{615\pi^2}{64}(\vec{n}_{12}\cdot\vec{V})^2 + 18v_1^2 + \frac{123\pi^2}{64}V^2 + 33(\vec{v}_1\cdot\vec{v}_2) - \frac{33}{2}v_2^2\bigg\} \\
&+ \frac{m_1^2}{r_{12}^2}\bigg\{-\frac{2069}{8}(\vec{n}_{12}\cdot\vec{v}_1)^2 + 543(\vec{n}_{12}\cdot\vec{v}_1)(\vec{n}_{12}\cdot\vec{v}_2) - \frac{939}{4}(\vec{n}_{12}\cdot\vec{v}_2)^2 \\
&+ \frac{471}{8}v_1^2 - \frac{357}{4}(\vec{v}_1\cdot\vec{v}_2) + \frac{357}{8}v_2^2\bigg\} \\
&+ \frac{16m_2^3}{r_{12}^3} + \frac{m_1^2 m_2}{r_{12}^3}\left(\frac{547}{3} - \frac{41\pi^2}{16}\right) - \frac{13m_1^3}{12r_{12}^3} + \frac{m_1 m_2^2}{r_{12}^3}\left(\frac{545}{3} - \frac{41\pi^2}{16}\right), \qquad (3.57)
\end{aligned}
$$

and finally

$$
\begin{aligned}
\mathcal{T}_4 = &\frac{15}{2}(\vec{n}_{12}\cdot\vec{v}_1)(\vec{n}_{12}\cdot\vec{v}_2)^4 - \frac{45}{8}(\vec{n}_{12}\cdot\vec{v}_2)^5 - \frac{3}{2}(\vec{n}_{12}\cdot\vec{v}_2)^3 v_1^2 \\
&+ 6(\vec{n}_{12}\cdot\vec{v}_1)(\vec{n}_{12}\cdot\vec{v}_2)^2(\vec{v}_1\cdot\vec{v}_2) - 6(\vec{n}_{12}\cdot\vec{v}_2)^3(\vec{v}_1\cdot\vec{v}_2) \\
&- 2(\vec{n}_{12}\cdot\vec{v}_2)(\vec{v}_1\cdot\vec{v}_2)^2 - 12(\vec{n}_{12}\cdot\vec{v}_1)(\vec{n}_{12}\cdot\vec{v}_2)^2 v_2^2 + 12(\vec{n}_{12}\cdot\vec{v}_2)^3 v_2^2 \\
&+ (\vec{n}_{12}\cdot\vec{v}_2)v_1^2 v_2^2 - 4(\vec{n}_{12}\cdot\vec{v}_1)(\vec{v}_1\cdot\vec{v}_2)v_2^2 + 8(\vec{n}_{12}\cdot\vec{v}_2)(\vec{v}_1\cdot\vec{v}_2)v_2^2 \\
&+ 4(\vec{n}_{12}\cdot\vec{v}_1)v_2^4 - 7(\vec{n}_{12}\cdot\vec{v}_2)v_2^4 \\
&+ \frac{m_2}{r_{12}}\{-2(\vec{n}_{12}\cdot\vec{v}_1)^2(\vec{n}_{12}\cdot\vec{v}_2) + 8(\vec{n}_{12}\cdot\vec{v}_1)(\vec{n}_{12}\cdot\vec{v}_2)^2 + 2(\vec{n}_{12}\cdot\vec{v}_2)^3 \\
&+ 2(\vec{n}_{12}\cdot\vec{v}_1)(\vec{v}_1\cdot\vec{v}_2) + 4(\vec{n}_{12}\cdot\vec{v}_2)(\vec{v}_1\cdot\vec{v}_2) \\
&- 2(\vec{n}_{12}\cdot\vec{v}_1)v_2^2 - 4(\vec{n}_{12}\cdot\vec{v}_2)v_2^2\}
\end{aligned}
$$

$$
\begin{aligned}
&+\frac{m_1}{r_{12}}\left\{-\frac{243}{4}(\vec{n}_{12}\cdot\vec{v}_1)^3+\frac{565}{4}(\vec{n}_{12}\cdot\vec{v}_1)^2(\vec{n}_{12}\cdot\vec{v}_2)\right.\\
&-\frac{269}{4}(\vec{n}_{12}\cdot\vec{v}_1)(\vec{n}_{12}\cdot\vec{v}_2)^2-\frac{95}{12}(\vec{n}_{12}\cdot\vec{v}_2)^3+\frac{207}{8}(\vec{n}_{12}\cdot\vec{v}_1)v_1^2\\
&-\frac{137}{8}(\vec{n}_{12}\cdot\vec{v}_2)v_1^2-36(\vec{n}_{12}\cdot\vec{v}_1)(\vec{v}_1\cdot\vec{v}_2)+\frac{27}{4}(\vec{n}_{12}\cdot\vec{v}_2)(\vec{v}_1\cdot\vec{v}_2)\\
&\left.+\frac{81}{8}(\vec{n}_{12}\cdot\vec{v}_1)v_2^2+\frac{83}{8}(\vec{n}_{12}\cdot\vec{v}_2)v_2^2\right\}\\
&+\frac{m_2^2}{r_{12}^2}\left(4(\vec{n}_{12}\cdot\vec{v}_1)+5(\vec{n}_{12}\cdot\vec{v}_2)\right)\\
&+\frac{m_1m_2}{r_{12}^2}\left\{-\frac{307}{8}(\vec{n}_{12}\cdot\vec{v}_1)+\frac{479}{8}(\vec{n}_{12}\cdot\vec{v}_2)+\frac{123\pi^2}{32}(\vec{n}_{12}\cdot\vec{V})\right\}\\
&+\frac{m_1^2}{r_{12}^2}\left\{\frac{311}{4}(\vec{n}_{12}\cdot\vec{v}_1)-\frac{357}{4}(\vec{n}_{12}\cdot\vec{v}_2)\right\}. \qquad (3.58)
\end{aligned}
$$

We note some properties of these 3PN equations of motion. In the test-particle limit these 3PN equations of motion coincide with the geodesic equations for a test particle in the Schwarzschild space-time in harmonic coordinates (up to 3PN order). Suppose that star 1 is a test particle, star 2 is the source of the Schwarzschild field, and $\vec{v}_2=0$. Then the geodesic equations for star 1 become

$$
\begin{aligned}
a_1^i=&-\frac{m_2}{r_{12}^2}n_{12}^i+\epsilon^2\frac{m_2}{r_{12}^2}\left(-v_1^2n_{12}^i+\frac{4m_2}{r_{12}}n_{12}^i+4(\vec{n}_{12}\cdot\vec{v}_1)v_1^i\right)\\
&+\epsilon^4\frac{m_2^2}{r_{12}^3}\left(2(\vec{n}_{12}\cdot\vec{v}_1)^2n_{12}^i-9\frac{m_2}{r_{12}}n_{12}^i-2(\vec{n}_{12}\cdot\vec{v}_1)v_1^i\right)\\
&+\epsilon^6\frac{m_2^3}{r_{12}^4}\left(-(\vec{n}_{12}\cdot\vec{v}_1)^3n_{12}^i+\frac{16m_2}{r_{12}}n_{12}^i+4(\vec{n}_{12}\cdot\vec{v}_1)v_1^i\right)+O(\epsilon^7), \qquad (3.59)
\end{aligned}
$$

in the harmonic gauge. Thus we confirm by comparison that in the test-particle limit (3.54) coincides with the geodesic equations for a test particle in the Schwarzschild space-time up to 3PN order.

With the help of the formulas developed in Blanchet and Faye (2001*a*) the Lorentz invariance (in the post-Newtonian perturbative sense) of (3.54) can be verified. In addition the 3PN acceleration given here admits a conserved energy of the binary orbital motion (modulo the 2.5PN radiation reaction effect). The energy E of the binary associated with (3.54) is

$$
E=\frac{1}{2}m_1v_1^2-\frac{m_1m_2}{2r_{12}}+\epsilon^2E_2+\epsilon^4E_4+\epsilon^6E_6+(1\leftrightarrow 2)+O(\epsilon^7), \qquad (3.60)
$$

where E_2, E_4, E_6 are given by

$$E_2 = \frac{3}{8}m_1 v_1^4 + \frac{m_1^2 m_2}{2r_{12}^2} + \frac{m_1 m_2}{2r_{12}} \left\{ 3v_1^2 - \frac{7}{2}(\vec{v}_1 \cdot \vec{v}_2) - \frac{1}{2}(\vec{n}_{12} \cdot \vec{v}_1)(\vec{n}_{12} \cdot \vec{v}_2) \right\}, \quad (3.61)$$

$$\begin{aligned} E_4 = &\frac{5}{16}m_1 v_1^6 - \frac{m_1^3 m_2}{2r_{12}^3} - \frac{19 m_1^2 m_2^2}{8r_{12}^3} + \frac{m_1^2 m_2}{2r_{12}^2} \left\{ -3v_1^2 + \frac{7}{2}v_2^2 \right. \\ &\left. + \frac{29}{2}(\vec{n}_{12} \cdot \vec{v}_1)^2 - \frac{13}{2}(\vec{n}_{12} \cdot \vec{v}_1)(\vec{n}_{12} \cdot \vec{v}_2) + (\vec{n}_{12} \cdot \vec{v}_2)^2 \right\} \\ &+ \frac{m_1 m_2}{4r_{12}} \left\{ \frac{3}{2}(\vec{n}_{12} \cdot \vec{v}_1)^3(\vec{n}_{12} \cdot \vec{v}_2) + \frac{3}{4}(\vec{n}_{12} \cdot \vec{v}_1)^2(\vec{n}_{12} \cdot \vec{v}_2)^2 \right. \\ &- \frac{9}{2}(\vec{n}_{12} \cdot \vec{v}_1)(\vec{n}_{12} \cdot \vec{v}_2)v_1^2 - \frac{13}{2}(\vec{n}_{12} \cdot \vec{v}_2)^2 v_1^2 + \frac{21}{2}v_1^4 \\ &+ \frac{13}{2}(\vec{n}_{12} \cdot \vec{v}_1)^2(\vec{v}_1 \cdot \vec{v}_2) + 3(\vec{n}_{12} \cdot \vec{v}_1)(\vec{n}_{12} \cdot \vec{v}_2)(\vec{v}_1 \cdot \vec{v}_2) \\ &\left. - \frac{55}{2}v_1^2(\vec{v}_1 \cdot \vec{v}_2) + \frac{17}{2}(\vec{v}_1 \cdot \vec{v}_2)^2 + \frac{31}{4}v_1^2 v_2^2 \right\}, \end{aligned} \quad (3.62)$$

and

$$\begin{aligned} E_6 = &\frac{35}{128}m_1 v_1^8 + \frac{3m_1^4 m_2}{8r_{12}^4} + \frac{469 m_1^3 m_2^2}{18 r_{12}^4} + \frac{m_1^2 m_2^2}{2r_{12}^3}\mathcal{S}_1 + \frac{m_1^3 m_2}{2r_{12}^3}\mathcal{S}_2 \\ &+ \frac{m_1 m_2}{r_{12}}\mathcal{S}_3 + \frac{m_1^2 m_2}{r_{12}^2}\mathcal{S}_4. \end{aligned} \quad (3.63)$$

In the latter equation $\mathcal{S}_1$, $\mathcal{S}_2$, $\mathcal{S}_3$, and $\mathcal{S}_4$ are defined by

$$\begin{aligned} \mathcal{S}_1 = &\frac{547}{6}(\vec{n}_{12} \cdot \vec{v}_1)^2 - \frac{3115}{24}(\vec{n}_{12} \cdot \vec{v}_1)(\vec{n}_{12} \cdot \vec{v}_2) - \frac{123\pi^2}{32}(\vec{n}_{12} \cdot \vec{v}_1)(\vec{n}_{12} \cdot \vec{V}) \\ &- \frac{575}{9}v_1^2 + \frac{41\pi^2}{32}(\vec{V} \cdot \vec{v}_2) + \frac{4429}{72}(\vec{v}_1 \cdot \vec{v}_2), \end{aligned} \quad (3.64)$$

$$\begin{aligned} \mathcal{S}_2 = &-\frac{437}{4}(\vec{n}_{12} \cdot \vec{v}_1)^2 + \frac{317}{4}(\vec{n}_{12} \cdot \vec{v}_1)(\vec{n}_{12} \cdot \vec{v}_2) + 3(\vec{n}_{12} \cdot \vec{v}_2)^2 + \frac{301}{12}v_1^2 \\ &- \frac{337}{12}(\vec{v}_1 \cdot \vec{v}_2) + \frac{5}{2}v_2^2, \end{aligned} \quad (3.65)$$

$$\begin{aligned} \mathcal{S}_3 = &-\frac{5}{16}(\vec{n}_{12} \cdot \vec{v}_1)^5(\vec{n}_{12} \cdot \vec{v}_2) - \frac{5}{16}(\vec{n}_{12} \cdot \vec{v}_1)^4(\vec{n}_{12} \cdot \vec{v}_2)^2 \\ &- \frac{5}{32}(\vec{n}_{12} \cdot \vec{v}_1)^3(\vec{n}_{12} \cdot \vec{v}_2)^3 + \frac{19}{16}(\vec{n}_{12} \cdot \vec{v}_1)^3(\vec{n}_{12} \cdot \vec{v}_2)v_1^2 \\ &+ \frac{15}{16}(\vec{n}_{12} \cdot \vec{v}_1)^2(\vec{n}_{12} \cdot \vec{v}_2)^2 v_1^2 + \frac{3}{4}(\vec{n}_{12} \cdot \vec{v}_1)(\vec{n}_{12} \cdot \vec{v}_2)^3 v_1^2 \end{aligned}$$

$$
\begin{aligned}
&+\frac{19}{16}(\vec{n}_{12}\cdot\vec{v}_2)^4 v_1^2 - \frac{21}{16}(\vec{n}_{12}\cdot\vec{v}_1)(\vec{n}_{12}\cdot\vec{v}_2)v_1^4 - 2(\vec{n}_{12}\cdot\vec{v}_2)^2 v_1^4 \\
&+\frac{55}{16}v_1^6 - \frac{19}{16}(\vec{n}_{12}\cdot\vec{v}_1)^4(\vec{v}_1\cdot\vec{v}_2) - (\vec{n}_{12}\cdot\vec{v}_1)^3(\vec{n}_{12}\cdot\vec{v}_2)(\vec{v}_1\cdot\vec{v}_2) \\
&-\frac{15}{32}(\vec{n}_{12}\cdot\vec{v}_1)^2(\vec{n}_{12}\cdot\vec{v}_2)^2(\vec{v}_1\cdot\vec{v}_2) + \frac{45}{16}(\vec{n}_{12}\cdot\vec{v}_1)^2 v_1^2(\vec{v}_1\cdot\vec{v}_2) \\
&+\frac{5}{4}(\vec{n}_{12}\cdot\vec{v}_1)(\vec{n}_{12}\cdot\vec{v}_2)v_1^2(\vec{v}_1\cdot\vec{v}_2) + \frac{11}{4}(\vec{n}_{12}\cdot\vec{v}_2)^2 v_1^2(\vec{v}_1\cdot\vec{v}_2) \\
&-\frac{139}{16}v_1^4(\vec{v}_1\cdot\vec{v}_2) - \frac{3}{4}(\vec{n}_{12}\cdot\vec{v}_1)^2(\vec{v}_1\cdot\vec{v}_2)^2 + \frac{5}{16}(\vec{n}_{12}\cdot\vec{v}_1)(\vec{n}_{12}\cdot\vec{v}_2)(\vec{v}_1\cdot\vec{v}_2)^2 \\
&+\frac{41}{8}v_1^2(\vec{v}_1\cdot\vec{v}_2)^2 + \frac{1}{16}(\vec{v}_1\cdot\vec{v}_2)^3 - \frac{45}{16}(\vec{n}_{12}\cdot\vec{v}_1)^2 v_1^2 v_2^2 \\
&-\frac{23}{32}(\vec{n}_{12}\cdot\vec{v}_1)(\vec{n}_{12}\cdot\vec{v}_2)v_1^2 v_2^2 + \frac{79}{16}v_1^4 v_2^2 - \frac{161}{32}v_1^2 v_2^2(\vec{v}_1\cdot\vec{v}_2),
\end{aligned}
\tag{3.66}
$$

and

$$
\begin{aligned}
\mathcal{S}_4 = &-\frac{49}{8}(\vec{n}_{12}\cdot\vec{v}_1)^4 + \frac{75}{8}(\vec{n}_{12}\cdot\vec{v}_1)^3(\vec{n}_{12}\cdot\vec{v}_2) \\
&-\frac{187}{8}(\vec{n}_{12}\cdot\vec{v}_1)^2(\vec{n}_{12}\cdot\vec{v}_2)^2 + \frac{11}{2}v_1^4 + \frac{247}{24}(\vec{n}_{12}\cdot\vec{v}_1)(\vec{n}_{12}\cdot\vec{v}_2)^3 \\
&+\frac{49}{8}(\vec{n}_{12}\cdot\vec{v}_1)^2 v_1^2 + \frac{81}{8}(\vec{n}_{12}\cdot\vec{v}_1)(\vec{n}_{12}\cdot\vec{v}_2)v_1^2 - \frac{21}{4}(\vec{n}_{12}\cdot\vec{v}_2)^2 v_1^2 \\
&-\frac{15}{2}(\vec{n}_{12}\cdot\vec{v}_1)^2(\vec{v}_1\cdot\vec{v}_2) - \frac{3}{2}(\vec{n}_{12}\cdot\vec{v}_1)(\vec{n}_{12}\cdot\vec{v}_2)(\vec{v}_1\cdot\vec{v}_2) \\
&+\frac{21}{4}(\vec{n}_{12}\cdot\vec{v}_2)^2(\vec{v}_1\cdot\vec{v}_2) - 27v_1^2(\vec{v}_1\cdot\vec{v}_2) + \frac{55}{2}(\vec{v}_1\cdot\vec{v}_2)^2 \\
&+\frac{49}{4}(\vec{n}_{12}\cdot\vec{v}_1)^2 v_2^2 - \frac{27}{2}(\vec{n}_{12}\cdot\vec{v}_1)(\vec{n}_{12}\cdot\vec{v}_2)v_2^2 + \frac{3}{4}(\vec{n}_{12}\cdot\vec{v}_2)^2 v_2^2 \\
&+\frac{55}{4}v_1^2 v_2^2 - 28(\vec{v}_1\cdot\vec{v}_2)v_2^2 + \frac{135}{16}v_2^4.
\end{aligned}
\tag{3.67}
$$

With E given by (3.60) [compare with E computed in Blanchet and Faye (2001*b*)] one can verify directly that E is conserved as a consequence of the 3PN equations of motion (3.54).

It is important to note that (3.44), when written out in full, also gives the correct geodesic equations in the test-particle limit, is Lorentz invariant, and admits the conserved energy. These facts can be seen from the form of $a_1^i|_{\delta_{A\,\ln}}$ in (3.48); $a_1^i|_{\delta_{A\,\ln}}$ vanishes when $m_1 \to 0$, is Lorentz invariant up to 3PN order and energy conservation is unaffected by the definition of the dipole moments.

We have retained, during the calculation of the 3PN equations of motion, $\mathcal{R}$-dependent terms with positive powers of $\mathcal{R}$ or logarithms of $\mathcal{R}$. As stated following (2.74), it is a useful computational check to show that our equations of motion do not depend on $\mathcal{R}$. In fact we found that the $\mathcal{R}$-dependent terms cancel each other out in the final result. There is no need to employ a gauge transformation to remove such $\mathcal{R}$ dependence. For terms with negative powers of $\mathcal{R}$ we simply assume that those terms cancel out the $\mathcal{R}$-dependent terms from the far zone contribution. Indeed Pati and Will (2000, 2002), whose method we have adopted to compute the far zone contribution, have proved that all the $\mathcal{R}$-dependent terms cancel out between the far zone and the near zone contributions through all post-Newtonian orders.

3.8 Observations and comments

A 3PN iteration result was first reported by Jaranowski and Schäfer (1998*b*, 1998*a*). There, a 3PN ADM Hamiltonian in the ADM transverse traceless (ADMTT) gauge for two point masses expressed as two Dirac delta distributions was derived based on the ADM canonical approach (Ohta *et al.*, 1973; Schäfer, 1986). However it was found in Jaranowski and Schäfer (1998*b*, 1998*a*) that the regularization they had used caused one coefficient ω_{kinetic} to be undetermined in their framework. Moreover they later found another undetermined coefficient in their Hamiltonian which they called ω_{static} (Jaranowski and Schäfer, 1999). Origins of these two coefficients were attributed to some unsatisfactory features of the regularization they had used, such as violation of the Leibniz rule. The former coefficient, which appears as a numerical multiplier of the term that depends on the momenta of the point particles, was then fixed as $\omega_{\text{kinetic}} = 41/24$ by *a posteriori* imposing Poincaré invariance on their 3PN Hamiltonian (Damour *et al.*, 2000). As for the latter coefficient, Damour *et al.* (2001*a*) succeeded in fixing it as $\omega_{\text{static}} = 0$, using dimensional regularization, and found the same value of ω_{kinetic} as in (Damour *et al.*, 2000), which ensures Lorentz invariance of their Hamiltonian. Jaranowski and Schäfer (2000) have shown that the 3PN ADM Hamiltonian with $\omega_{\text{static}} = 0$ corresponds to the Brill–Lindquist initial data.

Blanchet and Faye have tackled the derivation of the 3PN equations of motion for two point masses, described with the use of two Dirac delta functions, in harmonic coordinates (Blanchet and Faye 2000*b*, 2001*b*) based on their previous work (Blanchet *et al.*, 1998). The divergences due to their use of Dirac delta functions were systematically regularized with the help of Lorentz invariant generalized Hadamard partie finie regularization. They have extended the notion of the Hadamard partie finie regularization to regularize divergent integrals and a singular function which does not permit a power-like expansion near its singularities (Blanchet and Faye, 2000*a*). Furthermore the regularization is carefully constructed in Blanchet and Faye (2001*a*) so that it respects Lorentz invariance. Their equations of motion respect the Lorentz invariance in the post-Newtonian perturbative sense and admit a conserved energy of orbital motion modulo the 2.5PN radiation reaction effect. They found however that there exists one, and only one, undetermined coefficient which they call λ. Interestingly the two groups independently derived a transformation between the two gauges and found that these two results coincide with each other when there exists a relation (Damour *et al.*, 2001*b*; de Andrade *et al.*, 2001)

$$\omega_{\text{static}} = -\frac{11}{3}\lambda - \frac{1987}{840}. \tag{3.68}$$

Therefore it is possible to fix the λ parameter from the result of Damour *et al.* (2001*a*) as

$$\lambda = -\frac{1987}{3080}. \tag{3.69}$$

However the applicability of mathematical regularization to the current problem is not a trivial issue but an assumption to be verified, or at least supported by convincing arguments. There is no argument such as the "dominant Schwarzschild condition" (Damour, 1983) at 3PN order.

By comparing Equation (3.54) with the Blanchet and Faye 3PN equations of motion (Blanchet and Faye, 2001*b*) we find the following relation:

$$m_1 \vec{a}_1^{\text{this work}} = m_1 \left(\vec{a}_1^{\text{BF}}\right)_\lambda = -\frac{1987}{3080} + m_1 \vec{a}_1|_{\delta_{A\,\ln}} + m_1 \vec{a}_1|_{\delta_{A,\text{BF}}}, \tag{3.70}$$

where $m_1 \vec{a}_1^{\text{this work}}$ is the 3PN acceleration given in (3.54), $\left(\vec{a}_1^{\text{BF}}\right)_{\lambda=-1987/3080}$ is the Blanchet and Faye 3PN acceleration with $\lambda = -1987/3080$, and $m_1 \vec{a}_1|_{\delta_{A\,\ln}}$ is given by (3.48) with ϵR_A replaced by r'_A for notational consistency with the Blanchet and Faye 3PN equations of motion. $m_1 \vec{a}_1|_{\delta_{A,\text{BF}}}$ is an acceleration induced by the following dipole moments of the stars:

$$\delta^i_{A,\text{BF}} = -\frac{3709}{1260} m_A^3 a_A^i. \tag{3.71}$$

We can compute $m_1 a_1^i|_{\delta_{A,\text{BF}}}$ by substituting $\delta^i_{A,\text{BF}}$ instead of $\delta^i_{A\Theta}$ into Equation (3.39). Thus by choosing the dipole moments as

$$D^i_{A\Theta,\text{BF}} = \epsilon^4 \delta^i_{A\Theta} - \epsilon^4 \delta^i_{A,\text{BF}}, \tag{3.72}$$

we have the 3PN equations of motion in completely the same form as $\left(\vec{a}_1^{\text{BF}}\right)_{\lambda=-1987/3080}$. In other words the 3PN equations of motion derived above agree with $\left(\vec{a}_1^{\text{BF}}\right)_{\lambda=-1987/3080}$ modulo the definition of the dipole moments (or equivalently, the coordinate transformation under the harmonic coordinates condition). In Itoh and Futamase (2003), arguments are given to support this conclusion and the value of λ was found to be $\lambda = -1987/3080$, which is perfectly consistent with the relation (3.68) and the result of Damour *et al.* (2001*a*) ($\omega_{\text{static}} = 0$). Recently the equations of motion for compact binaries in the 3.5PN approximation have been derived by Itoh (2010).

4 Two-body problem in general relativity

For a binary star system the periastron advance per orbit period is calculated, as well as the Shapiro time delay in radar arrival time due to the passage of the signal through the gravitational field of the binary. The effect of gravitational radiation reaction on the orbital period and on the shape of the orbit is derived by studying the perturbations of the osculating elements of the orbit and, alternatively, by using traditional energy and angular momentum balance arguments. The gravitational interaction of spin is also discussed.

4.1 Equations of motion for a binary system

We consider a binary star system with m_1 and m_2 the masses of the stars and $\boldsymbol{r_1}$ and $\boldsymbol{r_2}$ their position 3-vectors. The 3-velocity of each star is denoted by $\boldsymbol{v_1}$ and $\boldsymbol{v_2}$ and the corresponding 3-accelerations are taken to be $\boldsymbol{a_1}$ and $\boldsymbol{a_2}$. The relative position vector of m_2 to m_1 and its direction are described by the vectors

$$\boldsymbol{r} = \boldsymbol{r_2} - \boldsymbol{r_1} \qquad \text{and} \qquad \boldsymbol{n} = \frac{\boldsymbol{r}}{|\boldsymbol{r}|}. \tag{4.1}$$

We denote the relative 3-velocity of m_2 to m_1 by

$$\boldsymbol{V} = \boldsymbol{v_2} - \boldsymbol{v_1}. \tag{4.2}$$

The distance between the stars and their relative speed are thus given by $r = |\boldsymbol{r}|$ and $V = |\boldsymbol{V}|$ respectively. To the order of approximation required in this chapter the post-Newtonian equations of motion read, using (3.54),

$$\begin{aligned} m_1 \frac{dv_1^i}{dt} &= -\frac{m_1 m_2}{r^2} n_{12}^i \\ &+ \frac{m_1 m_2}{r^2} \left[n_{12}^i \left(-v_1^2 - 2v_2^2 + \frac{3}{2}(\boldsymbol{v}_2 \cdot \boldsymbol{n}_{12})^2 + 4(\boldsymbol{v}_1 \cdot \boldsymbol{v}_2) + \frac{5m_1}{r} + \frac{4m_2}{r} \right) \right. \\ &\left. + V^i \left(4(\boldsymbol{v}_1 \cdot \boldsymbol{n}_{12}) - 3(\boldsymbol{v}_2 \cdot \boldsymbol{n}_{12}) \right) \right], \end{aligned} \tag{4.3}$$

$$
\begin{aligned}
m_2 \frac{dv_2^i}{dt} &= -\frac{m_1 m_2}{r^2} n_{21}^i \\
&+ \frac{m_1 m_2}{r^2} \left[n_{21}^i \left(-v_2^2 - 2v_1^2 + \frac{3}{2}(\boldsymbol{v}_1 \cdot \boldsymbol{n}_{12})^2 + 4(\boldsymbol{v}_1 \cdot \boldsymbol{v}_2) + \frac{5m_2}{r} + \frac{4m_1}{r} \right) \right. \\
&\left. - V^i \left(4(\boldsymbol{v}_2 \cdot \boldsymbol{n}_{21}) - 3(\boldsymbol{v}_1 \cdot \boldsymbol{n}_{21}) \right) \right], \qquad (4.4)
\end{aligned}
$$

where $\boldsymbol{n}_{21} = -\boldsymbol{n}_{12} = \boldsymbol{n}$, $v_1 = |\boldsymbol{v}_1|$ and $v_2 = |\boldsymbol{v}_2|$. These equations of motion provide a basis for discussing the classical experimental tests of general relativity [see, for example, Will (1993, 2006)]. Periastron advance, light bending, and Shapiro time delay play central roles in the post-Newtonian tests carried out in the solar field. These general relativistic effects are also observed in the gravitational field of the Hulse–Taylor binary pulsar. For applications to gravitational wave astronomy the two-body equations of motion have been obtained up to the 3.5 post-Newtonian order.

4.2 Periastron advance

In standard textbooks on general relativity the periastron advance is derived for a test particle or planet orbiting the sun. The Schwarzschild space-time is taken as a model of the solar gravitational field and calculation of the periastron advance is carried out using the time-like geodesic equations for the world-line of the planet in this space-time. Such a test particle treatment is not valid however for a two-body system with comparable masses, such as a binary pulsar. For the two-body problem, writing the relative 3-acceleration as $\boldsymbol{a} = \boldsymbol{a_2} - \boldsymbol{a}_1$, the equations of motion above can be written in the effective one-body form as

$$
\begin{aligned}
\boldsymbol{a} = -\frac{M\boldsymbol{r}}{r^3} + \frac{M\boldsymbol{r}}{r^3} &\left(\frac{4M}{r} - V^2 + \frac{2\mu}{r} \right. \\
&\left. -3\frac{\mu}{M}V^2 + \frac{3\mu}{2M}(\boldsymbol{r} \cdot \boldsymbol{n})^2 + \frac{M(\boldsymbol{r} \cdot \boldsymbol{V})\boldsymbol{V}}{r^3} \left[4 - \frac{2\mu}{M} \right] \right), \qquad (4.5)
\end{aligned}
$$

where we define the total mass

$$M = m_1 + m_2, \qquad (4.6)$$

and the reduced mass

$$\mu = \frac{m_1 m_2}{m_1 + m_2}. \qquad (4.7)$$

In the Newtonian two-body problem we do not usually solve the equations of motion. Instead, since the motion takes place in a 2-plane, we make use of the existence of the two constants of the motion, the specific angular momentum J and the specific energy E. In the same way we can shorten our post-Newtonian computations by first writing $\boldsymbol{r} = (r \cos \phi, r \sin \phi)$, and thereby introducing polar coordinates, and by making use of two constants of the motion.

With a dot indicating differentiation with respect to time t the conserved specific angular momentum in the post-Newtonian approximation is given by

$$r^2\dot{\phi} = J\left[1 - \frac{(M-3\mu)}{2\mu^2 M}\left(p_r^2 + \frac{\mu^2 J^2}{r^2}\right) - \frac{(3M+\mu)}{r}\right], \tag{4.8}$$

while the radial momentum p_r is related to the radial velocity $\dot{r}$ by

$$p_r = \mu\dot{r}\left[1 + \frac{M-3\mu}{2M}\left(\dot{r}^2 + \frac{J^2}{r^2}\right) + \frac{3M+2\mu}{r}\right]. \tag{4.9}$$

The conserved specific energy takes the form

$$E = E_N + E_{1PN}, \tag{4.10}$$

where the indicated Newtonian and post-Newtonian contributions are given by

$$E_N = \frac{1}{2}\dot{r}^2 + \frac{J^2}{2r^2} - \frac{M}{r}, \tag{4.11}$$

and

$$E_{1PN} = \frac{(M-3\mu)}{2M}\dot{r}^4 + \frac{(M-3\mu)J^2}{2Mr^2}\dot{r}^2 + \frac{(2M+5\mu)}{2r}\dot{r}^2 - \frac{(2M-\mu)J^2}{r^3} + \frac{M(2M-3\mu)}{2r^2}, \tag{4.12}$$

respectively. These expressions can also be derived from the post-Newtonian Lagrangian (or Hamiltonian). In similar fashion to the Newtonian case we eliminate the time derivative by combining (4.8) and (4.12). In the course of this we arrive at an algebraic equation for $du/d\phi$, where $u = r^{-1}$, which can be simplified to give the orbit equation

$$\begin{aligned}\left(\frac{du}{d\phi}\right)^2 &= \frac{2E}{J^2}\left(1 + \frac{(M-3\mu)E}{2M}\right) + \left(\frac{2M}{J^2} + \frac{2(4M-3\mu)E}{h^2}\right)u \\ &\quad + \left(-1 + \frac{3(2M-\mu)E}{J^2}\right)u^2 + \mu\, u^3 \\ &\equiv F(u).\end{aligned} \tag{4.13}$$

We can integrate this equation to find the post-Newtonian orbit as

$$\phi = \pm\int \frac{du}{\sqrt{F(u)}}, \tag{4.14}$$

but this integral is not elementary. Instead therefore we solve the orbit equation iteratively by looking for a solution of the form,

$$u(\phi) = u_N + u_{1PN} + O(2PN), \tag{4.15}$$

where the leading term is the Newtonian solution given by the Kepler orbit

$$u_N = \frac{GM}{h^2}(1 + e\cos\phi), \tag{4.16}$$

with the orbital eccentricity

$$e = \sqrt{1 + \frac{2Eh^2}{G^2M^2}}. \tag{4.17}$$

We employ the initial condition that at $t = 0$ we have $\phi = 0$ and $du/d\phi = 0$. Straightforward but lengthy computations give the post-Newtonian solution as

$$\begin{aligned} u_{1PN} = & \frac{M^2}{J^4}\left(4M + (2M - \mu)e^2\right) \\ & + \frac{M^2}{8J^4 e}\left\{(9M + \mu) + (38M - 6\mu)e^2 + (M - 3\mu)e^4\right\}\cos\phi \\ & + \frac{\mu M^2 e^2}{2J^4}\sin^2\phi + \frac{3M^3 e}{J^4}\phi\sin\phi. \end{aligned} \tag{4.18}$$

The periastron occurs at $\phi = \phi_{\text{peri}} = 0$ and corresponds to a maximum value of u. On the other hand the apoastron, corresponding to a minimum value of u, occurs at $\phi = \phi_{\text{apo}}$ with

$$\phi_{\text{apo}} = \pi\left(1 + \frac{3M^2}{J^2}\right). \tag{4.19}$$

Hence the periastron advance per orbit period, denoted by δ, is given by

$$\frac{\delta}{2\pi} \equiv \frac{2\phi_{\text{apo}} - 2\pi}{2\pi} = \frac{3M^2}{J^2}, \tag{4.20}$$

which clearly includes the periastron advance for a test-particle orbit as a special case. Extensions of this computation to include the second post-Newtonian approximation have been carried out by Antonacopoulos and Tsoupakis (1979), Damour and Schäfer (1988), and Ohta and Kimura (1989*a*) [see also Ohta *et al.* (1973), and Ohta and Kimura (1988, 1989*b*) where the behavior of the center of mass at higher order is discussed].

Using the *effective* one-body Hamiltonian at the second post-Newtonian order the periastron advance was discussed by Antonacopoulos and Tsoupakis (1979) and was correctly derived by Damour and Schäfer (1988) and independently by Ohta and Kimura (1989*a*). The Hamiltonian in this case takes the form,

$$\begin{aligned} H(r,\phi,p_r,p_\phi) = & \frac{1}{2\mu}\left(p_r^2 + \frac{p_\phi^2}{r^2}\right) - \frac{\mu M}{r} - \frac{(M - 3\mu)}{8\mu^3 M}\left(p_r^2 + \frac{p_\phi^2}{r^2}\right)^2 + \frac{\mu M^2}{2r^2} \\ & - \frac{1}{2\mu r}\left[(3M + \mu)\left(p_r^2 + \frac{p_\phi^2}{r^2}\right) + \mu p_r^2\right] \\ & + \frac{(M^2 - 5\mu M + 5\mu^2)}{16\mu^5 M^2}\left(p_r^2 + \frac{p_\phi^2}{r^2}\right)^3 - \frac{\mu M^2(M + 3\mu)}{4r^3} \\ & + \frac{M}{2\mu r^2}\left[(5M + 8\mu)\left(p_r^2 + \frac{p_\phi^2}{r^2}\right) + 3\mu p_r^2\right] \end{aligned}$$

$$+\frac{1}{8\mu^3 Mr}\left[(5M^2-20\mu M-3\mu^2)\left(p_r^2+\frac{p_\phi^2}{r^2}\right)^2-2\mu^2\left(p_r^2+\frac{p_\phi^2}{r^2}\right)p_r^2\right.$$
$$\left.-3\mu^2 p_r^4\right]. \tag{4.21}$$

Hamilton's canonical equation $\dot{\phi}=\partial H/\partial p_\phi$ leads to

$$\dot{\phi}=\frac{\mu J}{r^2}\left\{\left[\frac{1}{\mu}-\frac{M-3\mu}{2\mu^3 M}\left(p_r^2+\frac{\mu^2 J^2}{r^2}\right)-\frac{3M+\mu}{\mu r}\right]\right.$$
$$+\frac{3(M^2-5\mu M+5mu^2)}{8\mu^5 M^2}\left(p_r^2+\frac{\mu^2 J^2}{r^2}\right)^2+\frac{M(5M+8\mu)}{\mu r^2}$$
$$\left.+\frac{1}{2\mu^3 Mr}\left[(5M^2-20\mu M-3\mu^2)\left(p_r^2+\frac{\mu^2 J^2}{r^2}\right)-\mu^2 p_r^2\right]\right\}. \tag{4.22}$$

Hamilton's canonical equation $\dot{r}=\partial H/\partial p_r$ at the second post-Newtonian order yields

$$p_r=\dot{r}\left\{\mu+\frac{\mu(M-3\mu)}{2M}\left(\dot{r}^2+\frac{J^2}{r^2}\right)+\frac{\mu}{r}(3M+2\mu)\right.$$
$$-\frac{(M-3\mu)^2 J^2}{2\mu M^2 r^2}\left(\dot{r}^2+\frac{J^2}{r^2}\right)-\frac{(M-3\mu)(3M+\mu)J^2}{\mu M r^3}$$
$$+\frac{3\mu(M^2-7\mu M+13\mu^2)}{8M^2}\left(\dot{r}^2+\frac{J^2}{r^2}\right)^2+\frac{\mu(4M^2+\mu M+4\mu^2)}{r^2}$$
$$\left.+\frac{\mu}{2Mr}\left[(7M^2-10\mu M-14\mu^2)\left(\dot{r}^2+\frac{J^2}{r^2}\right)+2\mu(M-\mu)\dot{r}^2\right]\right\}. \tag{4.23}$$

The conserved specific energy up to the second post-Newtonian order has the form,

$$E=E_N+E_{1PN}+E_{2PN}, \tag{4.24}$$

where the 2PN term is given by

$$E_{2PN}=\frac{(51M^2-18\mu M+18\mu^2)J^2}{12r^4}-\frac{M(9M^2-15\mu M+24\mu^2)}{12r^3}$$
$$+\frac{(4M^2-27\mu M+48\mu^2)}{8M^2}\dot{r}^6+\frac{(2M^2-15\mu M+24\mu^2)J^2}{4M^2r^2}\dot{r}^4$$
$$-\frac{(3\mu M-12\mu^2)J^4}{8M^2r^4}\dot{r}^2+\frac{(24M^2-17\mu M-99\mu^2)}{8Mr}\dot{r}^4$$
$$+\frac{(4\mu M-8\mu^2)J^2}{Mr^3}\dot{r}^2+\frac{(3M^2+10\mu M+32\mu^2)}{4r^2}\dot{r}^2$$
$$-\frac{(M-3\mu)}{2M}\left\{\left(\frac{1}{2}\dot{r}^2+\frac{J^2}{2r^2}-\frac{M}{r}\right)^2\right.$$

$$+2\left(\frac{1}{2}\dot{r}^2+\frac{J^2}{2r^2}-\frac{M}{r}\right)$$
$$\times\left[\frac{(M-3\mu)}{2M}\dot{r}^4+\frac{(M-3\mu)J^2}{2Mr^2}\dot{r}^2+\frac{(2M+5\mu)}{2r}\dot{r}^2\right.$$
$$\left.-\frac{(2M-\mu)J^2}{r^3}+\frac{M(2M-3\mu)}{2r^2}\right]\Bigg\}$$
$$+\frac{(M^2-5\mu M+5\mu^2)}{2M^2}\left(\frac{1}{2}\dot{r}^2+\frac{J^2}{2r^2}-\frac{M}{r}\right)^3-\frac{7\mu J^4}{8r^5}. \tag{4.25}$$

With these equations we can eliminate the time-derivative terms as before and arrive at the orbit equation

$$\begin{aligned}\left(\frac{du}{d\phi}\right)^2 &= \frac{2E}{J^2}\left(1+\frac{(M-3\mu)E}{2M}-\frac{(M-4\mu)E^2}{2\mu M^2}\right)\\ &+\left(\frac{2M}{J^2}+\frac{2(4M-3\mu)E}{h^2}+\frac{(4M^2-10\mu M+4\mu^2)E^2}{mJ^2}\right)u\\ &+\left(-1+\frac{3(2M-\mu)E}{J^2}+\frac{(15M^2-24\mu M-6\mu^2)E)}{J^2}\right)u^2\\ &+\left(\mu+\frac{(M+7\mu)E}{M}+\frac{m(17M^2-27\mu M-12\mu^2)}{J^2}\right)u^3\\ &+\mu(M+8\mu)u^4-\frac{9\mu^2J^2}{M}u^5\\ &\equiv F(u).\end{aligned} \tag{4.26}$$

This equation is solved approximately by expanding $u(\phi)$ in the post-Newtonian sense (4.15). The differential equation for the second post-Newtonian order term takes the form

$$\begin{aligned}\sin\phi\,\frac{du_{2PN}}{d\phi}-\cos\phi\,u_{2PN} &= a_0+a_1\cos\phi+a_2\sin^2\phi+a_3\sin^2\phi\cos\phi\\ &+a_4\sin^2\phi\cos^2\phi+a_5\sin^4\phi\cos\phi\\ &+b_1(\phi^2-2\phi\sin\phi\cos\phi)+b_2\phi\sin^3\phi.\end{aligned} \tag{4.27}$$

The coefficients here are defined by,

$$\begin{aligned}a_0 = &-\frac{M^5}{8J^6e^3}\Big(8<53,-23,1>+2<413,-205,14>(e^2-1)\\ &+<448,-277,33>(e^2-1)^2+\frac{1}{2}<102,-100,29>(e^2-1)^3\\ &-\frac{1}{16}<1,2,-23>(e^2-1)^4\Big),\end{aligned} \tag{4.28}$$

$$a_1 = -\frac{M^5}{8J^6e^2}\Big(8 < 53, -23, 1 > +2 < 307, -159, 12 > (e^2 - 1)$$
$$+ < 194, -141, 22 > (e^2 - 1)^2 + < 4, -7, 6 > (e^2 - 1)^3\Big), \tag{4.29}$$

$$a_2 = \frac{M^5}{8J^6e}\Big(2 < 141, -43, 4 > +16 < 18, -7, 1 > (e^2 - 1)$$
$$+ < 33, -23, 8 > (e^2 - 1)^2\Big), \tag{4.30}$$

$$a_3 = \frac{M^5}{8J^6}\Big(< 34, 10, -12 > + < 34, 4, -20 > (e^2 - 1)$$
$$+ < 0, 3, -7 > (e^2 - 1)^2\Big), \tag{4.31}$$

$$a_4 = \frac{M^5e^3}{2J^6} < 0, 1, -4 >, \tag{4.32}$$

$$a_5 = \frac{M^5e^4}{8J^6} < 0, 0, 9 >, \tag{4.33}$$

$$b_1 = \frac{9M^5e}{2J^6}, \tag{4.34}$$

$$b_2 = \frac{3\mu M^4e^2}{J^6}, \tag{4.35}$$

where the symbol $< a, b, c >$ is defined by

$$< a, b, c > \equiv a + b\left(\frac{\mu}{M}\right) + c\left(\frac{\mu}{M}\right)^2. \tag{4.36}$$

The term contributing to the equation of the orbit (4.15) at the second post-Newtonian order is

$$u_{2PN} = -a_1 - a_0\cos\phi + (a_3 + b_2)\sin^2\phi + \frac{1}{2}a_4\sin^2\phi\cos\phi + \frac{1}{3}a_5\sin^4\phi$$
$$+ \left(a_2 + \frac{1}{2}a_4\right)\phi\sin\phi - b_2\phi\sin\phi\cos\phi - b_1\phi^2\cos\phi. \tag{4.37}$$

It follows again from the assumed initial condition that the periastron occurs at $\phi = \phi_{\rm peri} = 0$. On the other hand the apoastron occurs at $\phi = \phi_{\rm apo}$ where $du/d\phi = 0$. Hence we obtain

$$\phi_{\rm apo} = \pi\left[1 + \frac{3M^2}{J^2} + \frac{M^3}{4J^2}\left\{(90M - 24\mu) + (15M - 6\mu)\,e^2\right\}\right]. \tag{4.38}$$

Therefore the more accurate than (4.20) periastron advance per orbit period is given by

$$\frac{\delta}{2\pi} = \frac{3M^2}{J^2} + \frac{M^3}{4J^4}\left\{(90M - 24\mu) + (15M - 6\mu)\,e^2\right\}. \tag{4.39}$$

From this we recover the periastron advance for a test-particle orbit in the Schwarzschild field by taking the limit $\mu/M \to 0$:

$$\left(\frac{\delta}{2\pi}\right)_{\mathrm{Sch}} = \frac{3M^2}{J^2} + \frac{15M^4}{4J^4}(6+e^2). \tag{4.40}$$

Finally we consider the orbital period. To calculate this we use the equation of the orbit given by (4.15) with the second post-Newtonian term given by (4.37). Also with $\dot{r}$ in (4.23) we can obtain an expression for $\dot{\phi}$ using $\dot{r} = \dot{\phi}(dr/d\phi)$. Then the orbital period is given by the integral

$$P = \int_0^P dt = 2\int_0^{\phi_{\mathrm{apo}}} \left(\dot{\phi}\right)^{-1} d\phi. \tag{4.41}$$

Up to the second post-Newtonian order we find that

$$P = \frac{2\pi J^3}{M^2(1-e^2)^{3/2}}\left[1 + \frac{M(15M-\mu)}{8J^2}(1-e^2)\right.$$
$$\left. + \frac{M^2}{8J^4}\left\{(60M^2 - 24\mu M)(1-e^2)^{3/2} - \frac{1}{16}(105M^2 + 90\mu M + 9\mu^2)(1-e^2)^2\right\}\right]. \tag{4.42}$$

4.3 Light-like signals: Shapiro time delay

The first prediction of general relativity to be tested experimentally was the deflection of a photon path in the solar gravitational field. The two-body equations of motion that we are examining in this chapter can be derived from a Lagrangian. This Lagrangian can in turn be used to calculate the deflection of photon paths in the field of the two bodies. However the bending angle by the binary pulsar is too small to detect. On the other hand, the effect of the gravitational field of the binary on radar signal arrival time delay can be observed and has been measured. This so-called Shapiro time delay was originally investigated in order to test general relativity by observing the radar arrival time delay due to the passage of the signal through the solar gravitational field. Because of the technical simplicity of the derivation standard textbooks of general relativity discuss the Shapiro time delay by employing the Schwarzschild solution of Einstein's field equations in the first post-Newtonian approximation. This treatment is not adequate for compact binary systems. On the other hand the Shapiro time delay provides a test of general relativity using the binary pulsar and, in addition, yields an invaluable opportunity to measure the mass of the pulsar and that of the unseen companion.

Soon after the discovery of the Hulse–Taylor binary pulsar the importance of the time delay measurement for this system was realized. The arrival time delay due to a partial post-Newtonian effect was derived first by Wheeler (1975). Blandford and Teukolsky (1975) discussed in detail the arrival time delay as a tool for measuring the mass of PSR 1913+17, by using the short-time post-Newtonian effects, consisting of the gravitational redshift and the relativistic transverse Doppler shift of the observed pulse. Incorporating the Shapiro time delay in their earlier analysis Blandford and Teukolsky (1976*b*) corrected an error in the formula derived by Wheeler. Using this formula Epstein (1977*b*, 1979) pointed out the feasibility of determining the post-Keplerian parameters by using the full post-Newtonian timing effects.

The line-element, in the first post-Newtonian approximation, in a coordinate system with origin at the center of mass of the binary has the form [see Blandford and Teukolsky (1975, 1976*b*)]

$$ds^2 = -(1 + 2\Phi)dt^2 + (1 - 2\Phi)(dx^2 + dy^2 + dz^2), \tag{4.43}$$

where the potential Φ is given by

$$\Phi(\boldsymbol{x}, t) = \Phi_1 + \Phi_2 = -\frac{m_1}{|\boldsymbol{x} - \boldsymbol{r}_1|} - \frac{m_2}{|\boldsymbol{x} - \boldsymbol{r}_2|}. \tag{4.44}$$

The orbital plane of the binary system can be chosen as the equatorial plane of a polar coordinate system related to x, y, and z with the x-axis intersecting the binary orbit at the pericenter. The (r, θ, ϕ) coordinates of the pulsar and companion are $\boldsymbol{r}_1 = (r_1, \pi/2, \phi)$ and $\boldsymbol{r}_2 = (r_1, \pi/2, \phi + \pi)$, where ϕ is the true anomaly and r_1 and r_2 are, in the lowest approximation, ellipses about the mass center. We thus have

$$r_1 = \frac{m_2}{M} r, \qquad r_2 = \frac{m_1}{M} r \qquad \text{and} \qquad r = \frac{a(1 - e^2)}{1 + e\cos\phi}, \tag{4.45}$$

where a denotes the semi-major axis of the relative orbit and e the orbital eccentricity. The proper time τ_p measured by a clock co-moving with the pulsar is related to the coordinate time t via the line-element above by

$$(d\tau_p)^2 = -ds^2 = dt^2\left(1 + 2\Phi - v_1^2\right). \tag{4.46}$$

From Newtonian theory we have

$$v_1^2 = \frac{m_2^2}{M}\left(\frac{2}{r} - \frac{1}{a}\right). \tag{4.47}$$

Substituting this into (4.46), we find that

$$\frac{d\tau_p}{dt} = 1 - \frac{m_2}{r} - \frac{m_2^2}{M}\frac{1}{r}, \tag{4.48}$$

where we ignore constant terms that can be absorbed into the (non-observable) intrinsic frequency. It should be noted that $\Phi(r_1)$ is not really infinite. The clock is at the emitting radius of the pulsar and the term of $\Phi(r_1)$ merely contributes a constant gravitational redshift.

What we have to do now is to relate the distance r to the observable quantities. For this purpose we use the Kepler equation because we assume the Kepler orbit. The Kepler equation is written as

$$u - e\sin u = \frac{2\pi t}{T} + \sigma, \tag{4.49}$$

where u denotes the eccentric anomaly, T is the orbital period, and the constant σ is related to the periastron passage time. Using the equation for an elliptic orbit,

$$r = \frac{a}{1 - e\cos u}, \tag{4.50}$$

we integrate (4.48) and find that

$$\tau_p = t - \frac{m_2(m_1 + 2m_2)}{aM}\frac{Tu}{2\pi} = t - \frac{m_2(m_1 + 2m_2)}{aM}\frac{Te}{2\pi}\sin u, \tag{4.51}$$

where we have dropped constant σ in (4.49) by using the freedom of choice of the initial time and we have ignored an overall multiplicative constant as well. From the line-element (4.43) we obtain

$$\left|\frac{d\boldsymbol{x}}{dt}\right| = 1 + 2\Phi, \tag{4.52}$$

which is integrated to the required accuracy as

$$t_{\mathrm{arr}} - t_{\mathrm{em}} = \int_{\boldsymbol{r}_1(t_{\mathrm{em}})}^{\boldsymbol{r}_e(t_{\mathrm{arr}})} (1 - 2\Phi)|d\boldsymbol{x}|. \tag{4.53}$$

Here t_{arr} is the coordinate time of the pulse arrival, t_{em} the the coordinate time of emission and $\boldsymbol{r}_e$ the position vector of the earth with respect to the orbit barycenter. The integral of Φ is crucial for the general relativistic time delay across the photon orbit. In doing the integration we can take the pulsar and companion to be stationary at the leading order and we can ignore Φ_1, which simply adds a constant term. Then

$$-2\int_{\boldsymbol{r}_1}^{\boldsymbol{r}_e} \Phi|d\boldsymbol{x}| = 2m_2\int_{t_{\mathrm{em}}}^{t_{\mathrm{arr}}} \frac{dt}{|\boldsymbol{x}(t) - \boldsymbol{r}_2(t_{\mathrm{em}})|}. \tag{4.54}$$

Since we have at the zeroth order, namely in Euclidean space,

$$\boldsymbol{x}(t) = \boldsymbol{r}(t_{\mathrm{em}}) + \frac{t - t_{\mathrm{em}}}{t_{\mathrm{arr}} - t_{\mathrm{em}}}\left(\boldsymbol{r}_e(t_{\mathrm{arr}}) - \boldsymbol{r}_1(t_{\mathrm{em}})\right), \tag{4.55}$$

the integral becomes

$$2m_2\frac{(t_{\mathrm{arr}} - t_{\mathrm{em}})}{|\boldsymbol{r}_e - \boldsymbol{r}_1|}\ln\frac{|\boldsymbol{r}_e - \boldsymbol{r}_1||\boldsymbol{r} + \boldsymbol{r}_e - \boldsymbol{r}_1| + |\boldsymbol{r}_e - \boldsymbol{r}_1|^2 + \boldsymbol{r}\cdot(\boldsymbol{r}_e - \boldsymbol{r}_1)}{|\boldsymbol{r}_e - \boldsymbol{r}_1|r + \boldsymbol{r}\cdot(\boldsymbol{r}_e - \boldsymbol{r}_1)}. \tag{4.56}$$

In first approximation we have $t_{\mathrm{arr}} - t_{\mathrm{em}} = |\boldsymbol{r}_e - \boldsymbol{r}_1|$. Since the pulsar distance from the companion is much larger than the pulsar radius ($r_e \gg r_1$) the expression (4.56) reduces to

$$2m_2\ln\left(\frac{2r_e}{r + \boldsymbol{r}\cdot\boldsymbol{n}}\right), \tag{4.57}$$

where $\boldsymbol{n} \equiv \boldsymbol{r}_e/r_e$ is a unit vector pointing from the pulsar to the earth. The constant term $\ln(2r_e)$ can be ignored. The term $\boldsymbol{r}\cdot\boldsymbol{n}$ is equal to $-r\sin i\,\sin(\omega + \phi)$, where i is the inclination angle of the orbital plane and ω is the angular distance of the pericenter from the line of nodes measured in the direction of orbital motion. Hence the time-dependent part of the integral (4.57) is rewritten as

$$2m_2 \ln\left(\frac{1+e\cos\phi}{1-\sin i \sin(\omega+\phi)}\right). \tag{4.58}$$

Therefore we arrive at

$$t_{\rm arr} - t_{\rm em} = |\boldsymbol{r}_e(t_{\rm arr}) - \boldsymbol{r}_1(t_{\rm em})| + 2m_2 \ln\left(\frac{1+e\cos\phi}{1-\sin i \sin(\omega+\phi)}\right). \tag{4.59}$$

Now writing

$$|\boldsymbol{r}_e - \boldsymbol{r}_1| \approx r_b + (\boldsymbol{r}_b - \boldsymbol{r}_1)\cdot \boldsymbol{n}, \tag{4.60}$$

and substituting this into (4.59) we obtain

$$t_{\rm arr} = t_{\rm em} + r_b + (\boldsymbol{r}_b - \boldsymbol{r}_1) + 2m_2 \ln\left(\frac{1+e\cos\phi}{1-\sin i \sin(\omega+\phi)}\right). \tag{4.61}$$

Restoring units in which the gravitational constant G and the speed of light in a vacuum c are not both equal to unity we see that the Shapiro time delay is characterized by two parameters r and s with

$$r = \frac{Gm_2}{c^3}, \tag{4.62}$$

$$s = \sin i = \frac{a_1 \sin i}{m_2}\left(\frac{T}{2\pi}\right)^{-2/3} G^{-1/3} M^{2/3}. \tag{4.63}$$

The $O(c^{-2})$ correction to the Shapiro time delay has been derived by Kopeikin and Schäfer (1999). However the second post-Newtonian effect is too small to be detected by current instruments.

4.4 Orbital period decay via gravitational radiation

The success in explaining the orbital period decay that has been observed for the binary pulsar PSR1913+16 justifies the theory of general relativity at the level of approximation beyond the first post-Newtonian order. The standard textbooks of general relativity discuss the orbital period decay due to gravitational wave emission using balance arguments in which the energy radiated in the form of gravitational waves equals the source energy loss, and angular momentum carried away by gravitational radiation equals the source angular momentum loss. The balance argument is of great importance in general relativity but one has to validate such an argument from first principles by using Einstein's field equations. In spite of the usefulness of the balance argument it provides very little information concerning the precise position and velocity of each body. For this one has to solve *all* equations and not just a partial set of equations such as the energy balance equation. To achieve this we follow the practice adopted in celestial mechanics treatments [see for example Danby (1988) and Roy (1988)].

Writing the equations of motion for the Newtonian two-body problem in the form

$$\frac{d^2\boldsymbol{r}_1}{dt^2} = -\frac{Gm_1m_2(\boldsymbol{r}_1 - \boldsymbol{r}_2)}{|\boldsymbol{r}_1 - \boldsymbol{r}_2|^3}, \tag{4.64}$$

$$\frac{d^2\boldsymbol{r}_2}{dt^2} = -\frac{Gm_1m_2(\boldsymbol{r}_2 - \boldsymbol{r}_1)}{|\boldsymbol{r}_1 - \boldsymbol{r}_2|^3}, \tag{4.65}$$

we have seen that they can be reduced to the one-body form

$$\frac{d^2\boldsymbol{r}}{dt^2} = -\frac{Gm_1m_2\boldsymbol{r}}{|\boldsymbol{r}|^3}, \tag{4.66}$$

where $\boldsymbol{r} \equiv \boldsymbol{r}_2 - \boldsymbol{r}_1$ denotes the relative position vector whose components in Cartesian coordinates we write simply as x^i. The corresponding components of the relative 3-velocity $\dot{\boldsymbol{r}}$ are denoted v^i. The position $\boldsymbol{r}$ and velocity $\dot{\boldsymbol{r}}$ can be completely specified in terms of time and six arbitrary constants. For example one possibility is to use the following Keplerian elements (see Figure 4.1): a (the semi-major axis of the orbital ellipse), e (the orbital eccentricity), i (the inclination angle of the orbital plane), Ω (the longitude of the ascending node), ω (the argument of the pericenter), and t_0 (the periastron passage time). Instead of t_0 we shall use the mean anomaly at the epoch σ defined as

$$\sigma \equiv nt_0, \tag{4.67}$$

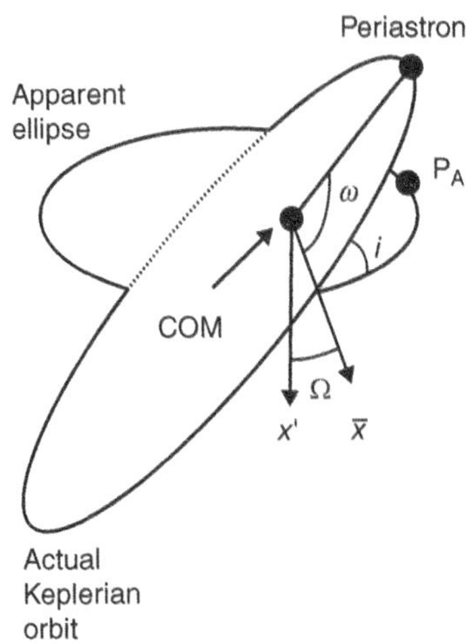

Fig. 4.1 Actual Keplerian orbit and apparent ellipse (projected onto the celestial sphere) in three-dimensional space. We show the inclination angle i, the argument of the periastron ω and the longitude of the ascending node Ω. These angles relate two coordinates (x', y') and $(\bar{x}, \bar{y})$, both of which have the origin as the common center of mass (denoted by COM in the figure). Here the x' axis is taken to lie along the major axis of the apparent ellipse, while the $\bar{x}$-axis is taken to lie along the direction of the ascending node.

where n denotes the mean motion as $\sqrt{GM/a^3} = 2\pi/T$. Combining the post-Newtonian measurements, the parameters of the Hulse–Taylor binary pulsar are fully specified. For example the pulsar and companion masses are derived as $m_p = 1.4414 \pm 0.0002$ and $m_c = 1.3867 \pm 0.0002$ solar masses respectively (Weisberg and Taylor, 2005).

4.5 Perturbations of contact elements

Useful connections between post-Newtonian corrections and celestial mechanics treatments are well described in Brumberg (1991). We shall therefore substantially follow this description although there are some notational differences. In particular the mean anomaly epoch σ is denoted l_0 by Brumberg. Also several alternative orbital parameterizations of the relativistic two-body problem are available [see Klioner and Kopeikin (1994) for instance].

Formally we can express the general solution of the two-body problem exactly as

$$x^i = f^i(c_1, c_2, c_3, c_4, c_5, c_6, t), \tag{4.68}$$

$$v^i = g^i(c_1, c_2, c_3, c_4, c_5, c_6, t), \tag{4.69}$$

where c_K for $K = 1$ to 6 denote six integrals of the motion and the lower index K labels each integral and is therefore not a tensorial index. For simplicity we shall however use the Einstein summation rule also for the non-tensorial capital letter indices with a repeated index to be summed over the range 1 to 6. We introduce perturbations by allowing c_K to be functions dependent on time. For the perturbed equations of motion we obtain

$$\frac{\partial f^i}{\partial c_K}\frac{dc_K}{dt} = 0, \tag{4.70}$$

$$\frac{\partial g^i}{\partial c_K}\frac{dc_K}{dt} = X^i. \tag{4.71}$$

By combining these we have

$$[c_K, c_J]\frac{dc_J}{dt} = X_j \frac{\partial x^j}{\partial c_K}, \tag{4.72}$$

where we define the Lagrange bracket as

$$[c_K, c_J] = \frac{\partial f^i}{\partial c_K}\frac{\partial g^i}{\partial c_J} - \frac{\partial g^i}{\partial c_K}\frac{\partial f^i}{\partial c_J} = \frac{\partial(x^i, \dot{x}^i)}{\partial(c_K, c_J)}. \tag{4.73}$$

The perturbing force $\boldsymbol{F}$ generally depends on $\boldsymbol{r}$, $\dot{\boldsymbol{r}}$, t (and possibly $\ddot{\boldsymbol{r}}$). If the perturbing force can be derived from a potential as

$$X_i = \frac{\partial R}{\partial x^i}, \tag{4.74}$$

then the above perturbed equations take the simpler form of

$$[c_K, c_J]\frac{dc_J}{dt} = \frac{\partial R}{\partial c_K}. \tag{4.75}$$

There are several options available in choosing the constants of motion for practical perturbation calculations. One useful choice is a, e, i, σ, ω, Ω. In this case the Lagrange brackets are

$$[\sigma, a] = \frac{1}{2}na, \tag{4.76}$$

$$[\omega, a] = \frac{1}{2}na\sqrt{1-e^2}, \tag{4.77}$$

$$[\omega, e] = -\frac{na^2 e}{\sqrt{1-e^2}}, \tag{4.78}$$

$$[\Omega, a] = \frac{1}{2}na\sqrt{1-e^2}\cos i, \tag{4.79}$$

$$[\Omega, e] = -na^2 e \cos\frac{i}{\sqrt{1-e^2}}, \tag{4.80}$$

$$[\Omega, i] = -na^2\sqrt{1-e^2}\sin\frac{i}{\sqrt{1-e^2}}. \tag{4.81}$$

In terms of this set of relations the perturbed equations of motion can be formally rewritten as

$$[\sigma, a]\frac{da}{dt} = \frac{\partial R}{\partial \sigma}, \tag{4.82}$$

$$[\omega, a]\frac{da}{dt} + [\omega, e]\frac{de}{dt} = \frac{\partial R}{\partial \omega}, \tag{4.83}$$

$$[\Omega, a]\frac{da}{dt} + [\Omega, e]\frac{de}{dt} + [\Omega, i]\frac{di}{dt} = \frac{\partial R}{\partial \Omega}, \tag{4.84}$$

$$[a, \sigma]\frac{d\sigma}{dt} + [a, \omega]\frac{d\omega}{dt} + [a, \Omega]\frac{d\Omega}{dt} = \frac{\partial R}{\partial a}, \tag{4.85}$$

$$[e, \omega]\frac{d\omega}{dt} + [e, \Omega]\frac{d\Omega}{dt} = \frac{\partial R}{\partial e}, \tag{4.86}$$

$$[i, \Omega]\frac{d\Omega}{dt} = \frac{\partial R}{\partial i}. \tag{4.87}$$

Substituting the explicit expressions (4.76)–(4.81) for the Lagrange brackets, these equations assume the form

$$\frac{da}{dt} = \frac{2}{na}\frac{\partial R}{\partial \sigma}, \tag{4.88}$$

$$\frac{de}{dt} = \frac{(1-e^2)}{na^2 e}\frac{\partial R}{\partial \sigma} - \frac{\sqrt{1-e^2}}{na^2 e}\frac{\partial R}{\partial \omega}, \tag{4.89}$$

$$\frac{di}{dt} = \frac{\cot i}{na^2\sqrt{1-e^2}}\frac{\partial R}{\partial \omega} - \frac{1}{na^2\sqrt{1-e^2}\sin i}\frac{\partial R}{\partial \Omega}, \tag{4.90}$$

$$\frac{d\sigma}{dt} = -\frac{2}{na}\frac{\partial R}{\partial a} - \frac{(1-e^2)}{na^2 e}\frac{\partial R}{\partial e}, \tag{4.91}$$

$$\frac{d\omega}{dt} = \frac{\sqrt{1-e^2}}{na^2 e}\frac{\partial R}{\partial e} - \frac{\cot i}{na^2\sqrt{1-e^2}}\frac{\partial R}{\partial i}, \tag{4.92}$$

$$\frac{d\Omega}{dt} = \frac{1}{na^2\sqrt{1-e^2}\sin i}\frac{\partial R}{\partial i}. \tag{4.93}$$

These six equations were derived first by Lagrange and thus are called the Lagrange planetary equations.

In most applications the perturbed equations above are very difficult to solve exactly. The most simple technique is to substitute into the right-hand sides of these equations the constant values for the elements and then obtain by integration their first-order perturbations. For the first-order secular perturbations of the constant elements the equations become

$$\frac{da}{dt} = \frac{2}{na}\frac{\partial <R>}{\partial \sigma}, \tag{4.94}$$

$$\frac{de}{dt} = \frac{(1-e^2)}{na^2 e}\frac{\partial <R>}{\partial \sigma} - \frac{\sqrt{1-e^2}}{na^2 e}\frac{\partial <R>}{\partial \omega}, \tag{4.95}$$

$$\frac{di}{dt} = \frac{\cot i}{na^2\sqrt{1-e^2}}\frac{\partial <R>}{\partial \omega} - \frac{1}{na^2\sqrt{1-e^2}\sin i}\frac{\partial <R>}{\partial \Omega}, \tag{4.96}$$

$$\frac{d\sigma}{dt} = -\frac{2}{na}\frac{\partial <R>}{\partial a} - \frac{(1-e^2)}{na^2 e}\frac{\partial <R>}{\partial e}, \tag{4.97}$$

$$\frac{d\omega}{dt} = \frac{\sqrt{1-e^2}}{na^2 e}\frac{\partial <R>}{\partial e} - \frac{\cot i}{na^2\sqrt{1-e^2}}\frac{\partial <R>}{\partial i}, \tag{4.98}$$

$$\frac{d\Omega}{dt} = \frac{1}{na^2\sqrt{1-e^2}\sin i}\frac{\partial <R>}{\partial i}, \tag{4.99}$$

where the brackets $<>$ denote averaging over the fast changing angular variables. In our case the perturbing force is small in the post-Newtonian sense. For $\boldsymbol{F} = 0$ we can solve the equations of motion and the position and velocity can be expressed in terms of time and six constants such as the Keplerian elements. It may be useful however to replace the Keplerian elements by vectorial elements, for example by the mutually orthogonal areal vector (specific angular momentum) $\boldsymbol{c}$ and Laplace–Runge–Lenz vector $\boldsymbol{e}$. These are defined by

$$\boldsymbol{c} = \boldsymbol{r} \times \dot{\boldsymbol{r}}, \tag{4.100}$$

and

$$\boldsymbol{e} = \left(\dot{\boldsymbol{r}}^2 - \frac{GM}{r}\right)\boldsymbol{r} - (\boldsymbol{r}\cdot\dot{\boldsymbol{r}})\dot{\boldsymbol{r}}, \tag{4.101}$$

respectively. It is convenient to introduce the triad of unit vectors that are directed along the line of nodes towards the ascending node of the orbit, transverse to the line of nodes in the orbital plane and transverse to the orbital plane, given respectively by

$$\boldsymbol{l} = (\cos\Omega, \sin\Omega, 0), \tag{4.102}$$

$$\boldsymbol{m} = (-\cos i \sin\Omega, \cos i \cos\Omega, \sin i), \tag{4.103}$$

$$\boldsymbol{k} = (\sin i \sin\Omega, -\sin i \cos\Omega, \cos i). \tag{4.104}$$

If we define $\boldsymbol{P}$ and $\boldsymbol{Q}$ by

$$\boldsymbol{P} = \boldsymbol{l}\cos\omega + \boldsymbol{m}\sin\omega, \tag{4.105}$$

$$\boldsymbol{Q} = -\boldsymbol{l}\sin\omega + \boldsymbol{m}\cos\omega, \tag{4.106}$$

then these are unit vectors in the orbital plane directed along the line of apsides towards the pericenter and transverse to the line of apsides respectively. By straightforward calculation one can show that

$$\mathbf{c} = \sqrt{GMa(1-e^2)}\,\boldsymbol{k}, \tag{4.107}$$

and

$$\mathbf{e} = GMe\,\boldsymbol{P}. \tag{4.108}$$

The position of the body in the orbital plane is described using coordinates consisting of the distance r and the argument of latitude u by

$$\boldsymbol{r} = r\,(\boldsymbol{l}\cos u + \boldsymbol{m}\sin u). \tag{4.109}$$

In addition we have

$$\dot{\boldsymbol{r}} = \sqrt{\frac{GM}{a(1-e^2)}}\,[-\boldsymbol{l}(\sin u + e\sin\omega) + \boldsymbol{m}(\cos u + e\cos\omega)]\,. \tag{4.110}$$

Alternatively we may use the true anomaly f which is the angular distance between the pericenter and the moving body. Then

$$u = f + \omega. \tag{4.111}$$

In terms of f the equation for the eccentric orbit reads

$$r = \frac{a(1-e^2)}{1+e\cos f}. \tag{4.112}$$

Then the position and velocity vectors are given by

$$\boldsymbol{r} = r\,(\boldsymbol{P}\cos f + \boldsymbol{Q}\sin f), \tag{4.113}$$

and

$$\dot{\boldsymbol{r}} = \sqrt{\frac{GM}{a(1-e^2)}}\,[-\boldsymbol{P}\sin f + \boldsymbol{Q}(\cos f + e)]\,. \tag{4.114}$$

The eccentric anomaly U is related to the true anomaly f via

$$\tan\frac{f}{2} = \sqrt{\frac{1+e}{1-e}}\tan\frac{U}{2}. \tag{4.115}$$

4.6 Perturbations of osculating elements

We have focused above on the case of a perturbing force (4.74) that can be derived from a disturbing function R. For more general cases such as a dissipative force this treatment cannot be used. In order to investigate such a general case we must now discuss perturbations of osculating elements.

The perturbed equations of motion have the general form

$$\frac{d^2\boldsymbol{r}}{dt^2} = -\frac{GM}{r^3}\boldsymbol{r} + \boldsymbol{F}. \tag{4.116}$$

Let S, T, W denote the components of the perturbing force per mass $\boldsymbol{F}$ along the vector $\boldsymbol{r}$, transverse to this vector and normal to the orbital plane respectively. Hence

$$S = \frac{\boldsymbol{r}\cdot\boldsymbol{F}}{r}, \tag{4.117}$$

$$T = \frac{(\boldsymbol{k}\cdot\boldsymbol{r})\cdot\boldsymbol{F}}{r}, \tag{4.118}$$

$$W = \boldsymbol{k}\cdot\boldsymbol{F}. \tag{4.119}$$

For the vectorial elements we have the perturbed equations,

$$\frac{d\boldsymbol{c}}{dt} = \boldsymbol{r}\times\boldsymbol{F}, \tag{4.120}$$

$$\frac{d\boldsymbol{e}}{dt} = 2(\dot{\boldsymbol{r}}\cdot\boldsymbol{F})\boldsymbol{r} - (\boldsymbol{r}\cdot\boldsymbol{F})\dot{\boldsymbol{r}} - (\boldsymbol{r}\cdot\dot{\boldsymbol{r}})\boldsymbol{F}. \tag{4.121}$$

These equations can be written more explicitly as

$$\frac{d\boldsymbol{c}}{dt} = \frac{na}{2\sqrt{1-e^2}}\frac{d}{dt}\{a(1-e^2)\}\boldsymbol{k} + na^2\sqrt{1-e^2}\left(\boldsymbol{l}\sin i\frac{d\Omega}{dt} - \boldsymbol{m}\frac{di}{dt}\right), \tag{4.122}$$

and

$$\begin{aligned}\frac{d\boldsymbol{e}}{dt} &= n^2a^3\frac{de}{dt}\boldsymbol{P}\\ &+n^2a^3e\left[\left(\frac{d\omega}{dt} + \cos i\frac{d\Omega}{dt}\right)\boldsymbol{Q} + \left(\sin\omega\frac{di}{dt} - \cos\omega\sin i\frac{d\Omega}{dt}\right)\boldsymbol{k}\right].\end{aligned} \tag{4.123}$$

The scalar product of $\dot{\boldsymbol{c}}$ with $\boldsymbol{k}$, $\boldsymbol{l}$, and $\boldsymbol{m}$ and the scalar product of $\dot{\boldsymbol{f}}$ with $\boldsymbol{P}$, $\boldsymbol{Q}$, and $\boldsymbol{k}$ give the following six equations:

$$\frac{da}{dt} = \frac{2}{n\sqrt{1-e^2}}\left(Se\sin f + T\frac{a\sqrt{1-e^2}}{r}\right), \tag{4.124}$$

$$\frac{de}{dt} = \frac{\sqrt{1-e^2}}{na}\left[S\sin f + T(\cos f + \cos U)\right], \tag{4.125}$$

$$\frac{di}{dt} = \frac{r\cos(f+\omega)}{na^2\sqrt{1-e^2}}W, \tag{4.126}$$

$$\frac{d\Omega}{dt} = \frac{r\sin(f+\omega)}{na^2\sqrt{1-e^2}\sin i}W, \tag{4.127}$$

$$\frac{d\omega}{dt} = -\cos i\frac{d\Omega}{dt} + \frac{\sqrt{1-e^2}}{nae}\left[-S\cos f + T\left(1+\frac{r}{a\sqrt{1-e^2}}\right)\sin f\right], \tag{4.128}$$

$$\frac{d\sigma}{dt} = -\sqrt{1-e^2}\left(\frac{d\omega}{dt} + \cos i\frac{d\Omega}{dt}\right) - S\frac{2r}{na^2}. \tag{4.129}$$

4.7 Perturbations due to gravitational radiation reaction

Up to the 2.5 post-Newtonian order the first body must satisfy equations of motion of the form

$$\frac{d^2x_1^i}{dt^2} = A_0^i(\boldsymbol{x}_1-\boldsymbol{x}_2) + A_2^i(\boldsymbol{x}_1-\boldsymbol{x}_2,\boldsymbol{v}_1,\boldsymbol{v}_2) + A_4^i(\boldsymbol{x}_1-\boldsymbol{x}_2,\boldsymbol{v}_1,\boldsymbol{v}_2) + A_5^i(\boldsymbol{x}_1-\boldsymbol{x}_2,\boldsymbol{v}_1-\boldsymbol{v}_2), \tag{4.130}$$

with similar equations of motion for the second body [see Damour and Deruelle (1981*a*, 1981*b*, 1981*c*), Kopeikin (1985), Barker and O'Connell (1975*a*, 1975*b*), and Futamase (1985*b*)]. The force per mass terms on the right-hand sides of these equations are given by

$$A_0^i = -\frac{Gm_2}{r^2}N^i, \tag{4.131}$$

$$A_2^i = \frac{Gm_2}{r^2}\left[N^i\left(-v_1^2 - 2v_2^2 + 4(\boldsymbol{v}_1\cdot\boldsymbol{v}_2) + \frac{3}{2}v_1^2(\boldsymbol{N}\cdot\boldsymbol{v}_2)^2 + \frac{G(5m_1+4m_2)}{r}\right) + v^i(4\boldsymbol{N}\cdot\boldsymbol{v}_1 - 3\boldsymbol{N}\cdot\boldsymbol{v}_2)\right], \tag{4.132}$$

$$A_4^i = \frac{4m_1m_2}{5r^3}\left[N^i(\boldsymbol{N}\cdot\boldsymbol{V})\left(-6\frac{m_1}{r} + \frac{52}{3}\frac{m_2}{r} + 3V^2\right) + V^i\left(2\frac{m_1}{r} - 8\frac{m_2}{r} - V^2\right)\right], \tag{4.133}$$

$$A_5^i = \frac{Gm_2}{r^2}N^i \left[-2v_2^4 + 4v_2^2(\boldsymbol{v}_1 \cdot \boldsymbol{v}_2) - 2(\boldsymbol{v}_1 \cdot \boldsymbol{v}_2)^2 + \frac{3}{2}v_1^2(\boldsymbol{N} \cdot \boldsymbol{v}_2)^2 \right.$$
$$+\frac{9}{2}v_2^2(\boldsymbol{N} \cdot \boldsymbol{v}_2)^2 - 6(\boldsymbol{v}_1 \cdot \boldsymbol{v}_2)(\boldsymbol{N} \cdot \boldsymbol{v}_2)^2 - \frac{15}{8}(\boldsymbol{N} \cdot \boldsymbol{v}_2)^4$$
$$+\frac{m_1}{r}\left(-\frac{15}{4}v_1^2 + \frac{5}{4}v_2^2 - \frac{5}{2}(\boldsymbol{v}_1 \cdot \boldsymbol{v}_2) + \frac{39}{2}(\boldsymbol{N} \cdot \boldsymbol{v}_1)^2 - 39(\boldsymbol{N} \cdot \boldsymbol{v}_1)(\boldsymbol{N} \cdot \boldsymbol{v}_2) \right.$$
$$\left. +\frac{17}{2}(\boldsymbol{N} \cdot \boldsymbol{v}_2)^2 \right) + \frac{m_2}{r}\left(4v_2^2 - 8(\boldsymbol{v}_1 \cdot \boldsymbol{v}_2) + 2(\boldsymbol{N} \cdot \boldsymbol{v}_1)^2 - 4(\boldsymbol{N} \cdot \boldsymbol{v}_1)(\boldsymbol{N} \cdot \boldsymbol{v}_2)\right.$$
$$\left.\left. +6(\boldsymbol{N} \cdot \boldsymbol{v}_2)^2\right) - \frac{57}{4}\frac{m_1^2}{r^2} - 9\frac{m_2^2}{r^2} - \frac{69}{2}\frac{m_1 m_2}{r^2} \right] + \frac{m_2}{r^2}V^i \left[v_1^2(\boldsymbol{N} \cdot \boldsymbol{v}_2) \right.$$
$$+4v_2^2(\boldsymbol{N} \cdot \boldsymbol{v}_1) - 5v_2^2(\boldsymbol{N} \cdot \boldsymbol{v}_2) - 4(\boldsymbol{v}_1 \cdot \boldsymbol{v}_2)(\boldsymbol{N} \cdot \boldsymbol{v}) - 6(\boldsymbol{N} \cdot \boldsymbol{v}_1)(\boldsymbol{N} \cdot \boldsymbol{v}_2)^2$$
$$+\frac{9}{2}(\boldsymbol{N} \cdot \boldsymbol{v}_2)^3 + \frac{m_1}{r}\left(-\frac{63}{4}(\boldsymbol{N} \cdot \boldsymbol{v}_1) + \frac{55}{4}(\boldsymbol{N} \cdot \boldsymbol{v}_2) \right)$$
$$\left. +\frac{m_2}{r}\left(-2(\boldsymbol{N} \cdot \boldsymbol{v}_1) - 2(\boldsymbol{N} \cdot \boldsymbol{v}_2)\right) \right]. \tag{4.134}$$

We now focus on the effect of the radiation reaction term A_4^i. Putting $F^i = A_4^i$ into (4.120) and (4.121) we obtain

$$\frac{d\boldsymbol{c}}{dt} = -\frac{4}{5}\frac{G^3 M m_1 m_2}{c^5 a^4}\frac{(8 + 7e^2)}{(1 - e^2)^{5/2}}\,\mathbf{c}, \tag{4.135}$$

and

$$\frac{d\boldsymbol{e}}{dt} = -\frac{1}{15}\frac{G^4 M^2 m_1 m_2}{c^5 a^4}\frac{(304 + 121e^2)e}{(1 - e^2)^{5/2}}\boldsymbol{P}. \tag{4.136}$$

Taking the scalar products of (4.135) with $\boldsymbol{l}$, $\boldsymbol{m}$, and $\boldsymbol{k}$ yields

$$\frac{di}{dt} = 0, \tag{4.137}$$

$$\frac{d\Omega}{dt} = 0, \tag{4.138}$$

$$\frac{d}{dt}\{a(1 - e^2)\} = -\frac{8}{5}\frac{G^3 M m_1 m_2}{c^5 a^3}\frac{(8 + 7e^2)}{(1 - e^2)^{3/2}}. \tag{4.139}$$

While the scalar products of (4.136) with $\boldsymbol{Q}$ and $\boldsymbol{P}$ give

$$\frac{d\omega}{dt} = 0, \tag{4.140}$$

$$\frac{de}{dt} = -\frac{1}{15}\frac{G^3 M m_1 m_2}{c^5 a^4}\frac{(304 + 121e^2)}{(1 - e^2)^{5/2}}e. \tag{4.141}$$

The first of these two equations tells us that *no* periastron advance is produced by the gravitational radiation reaction. The second of these two equations is an evolution equation for the orbital eccentricity implying that it gradually decreases owing to the radiation reaction. Substitution of (4.121) into (4.139) leads to

$$\frac{da}{dt} = -\frac{64}{5}\frac{G^4}{c^5}\frac{M^2\mu}{a^3(1-e^2)^{7/2}}\left(1+\frac{73}{24}e^2+\frac{37}{96}e^4\right), \tag{4.142}$$

which is an evolution equation for the orbital separation showing that the separation gradually decreases due to the radiation reaction.

4.8 Derivations by balance arguments

In the standard textbooks of general relativity balance arguments are used for discussion of the orbital shrinking rate due to radiation reaction effects. The stress–energy tensor corresponding to the gravitational waves is expressed as [see Isaacson (1968*a*, 1968*b*)]

$$T^{GW}_{\mu\nu} = \frac{c^4}{32\pi G}\left\langle \bar{h}_{\alpha\beta|\mu}\bar{h}^{\alpha\beta}{}_{|\mu} - \bar{h}_{\alpha\beta}{}^{|\alpha}\left(\bar{h}^{\beta}{}_{\mu|\nu}+\bar{h}^{\beta}{}_{\nu|\mu}\right) - \frac{1}{2}\bar{h}_{|\mu}\bar{h}_{|\nu}\right\rangle_{BHI}, \tag{4.143}$$

where the stroke denotes the covariant derivative with respect to the Riemannian connection calculated with the background metric (described below) and the bracket $<>_{BHI}$ denotes the Brill–Hartle–Isaacson averaging over several wavelengths. A derivation of this energy–momentum tensor, which we will briefly summarize here, can be found in Misner *et al.* (1973). The approach of Isaacson mentioned above was based on earlier work by Brill and Hartle (1964). The method is the *short-wavelength* approximation. We demand both $A \ll 1$ and $\lambda/R \ll 1$ where A is the typical amplitude of the waves, λ is the typical wavelength, and R is the typical radius of curvature of the background space-time. The full metric is decomposed into a background and a perturbation as

$$g_{\mu\nu} = g^{(B)}_{\mu\nu} + h_{\mu\nu}. \tag{4.144}$$

It follows that the Ricci tensor is expanded as

$$R_{\mu\nu} = R^{(B)}_{\mu\nu} + R^{(1)}_{\mu\nu}(h) + R^{(2)}_{\mu\nu}(h) + \cdots, \tag{4.145}$$

where the superscripts (1) and (2) mean linear and second order in $h_{\mu\nu}$ respectively. The perturbed metric $h_{\mu\nu}$ must satisfy

$$R^{(1)}_{\mu\nu}(h) = 0. \tag{4.146}$$

Hence we can rewrite Einstein's vacuum field equations in the *effective* form

$$R^{(B)}_{\mu\nu} - \frac{1}{2}R^{(B)}g^{(B)}_{\mu\nu} = 8\pi T^{GW}_{\mu\nu}, \tag{4.147}$$

where we define

$$R^{(B)} \equiv g^{\mu\nu(B)}R^{(B)}_{\mu\nu}, \tag{4.148}$$

and the *effective* stress–energy tensor for gravitational waves is defined by

$$-8\,\pi\, T^{GW}_{\mu\nu} = \left\langle R^{(2)}_{\mu\nu}(h) \right\rangle_{BHI} - \frac{1}{2} g^{(B)}_{\mu\nu} \left\langle R^{(2)}(h) \right\rangle_{BHI}. \tag{4.149}$$

Useful rules for manipulating quantities inside the averaging brackets are listed in Misner *et al.* (1973). A straightforward but lengthy calculation, using these rules and the field equation (4.146) for $h_{\mu\nu}$, leads to (4.143).

For our purposes we write the background metric tensor components in the form

$$g^{(B)}_{\mu\nu} = \eta_{\mu\nu} + h_{\mu\nu}, \tag{4.150}$$

where $\eta_{\mu\nu}$ is the Minkowskian metric tensor and $h_{\mu\nu}$ is a small perturbation. As usual it will be convenient to define

$$\bar{h}_{\mu\nu} = h_{\mu\nu} - \frac{1}{2} h\, \eta_{\mu\nu}, \tag{4.151}$$

with $h = \eta^{\mu\nu} h_{\mu\nu}$ and thus $\bar{h} = -h$. The stress–energy tensor for gravitational waves can be written as

$$T^{GW}_{\mu\nu} = \langle \bar{h}^{TT}_{jk,\mu} \bar{h}^{TT}_{jk,\nu} \rangle_{BHI}, \tag{4.152}$$

where TT denotes the transverse traceless part of $\bar{h}_{\mu\nu}$. At the quadrupole approximation order we find in particular that

$$T^{GW}_{00} = \frac{1}{8\pi r^2} \left\langle \frac{d^3}{dt^3} Q^{TT}_{jk}(t-r) \frac{d^3}{dt^3} Q^{TT}_{jk}(t-r) \right\rangle_{BHI} + O\left(\frac{1}{r^3}\right), \tag{4.153}$$

in the wave zone for which the distance from the source is much larger than the wavelength and the source size. Here Q_{jk} denotes the traceless part of the quadrupole moment of the source. From this we obtain the gravitational radiation power as

$$\begin{aligned} L_{GW} &= \frac{1}{8\pi} \oint \left\langle \frac{d^3}{dt^3} Q^{TT}_{jk} \frac{d^3}{dt^3} Q^{TT}_{jk} \right\rangle_{BHI} d\Omega \\ &= \frac{1}{8\pi} \left\langle \oint \left[\frac{d^3}{dt^3} Q_{jk} \frac{d^3}{dt^3} Q_{jk} - 2n^i \frac{d^3}{dt^3} Q_{ij} \frac{d^3}{dt^3} Q_{jk} n^k \right.\right. \\ &\qquad \left.\left. + \frac{1}{2} \left(n^i \frac{d^3}{dt^3} Q_{ij} n^j \right)^2 \right] d\Omega \right\rangle_{BHI} \\ &= \frac{1}{5} \left\langle \frac{d^3}{dt^3} Q_{jk} \frac{d^3}{dt^3} Q_{jk} \right\rangle_{BHI}, \end{aligned} \tag{4.154}$$

where the integration is taken over the unit 2-sphere and the last line is referred to in the literature as the quadrupole formula. In units for which the gravitational constant G and the speed of light in a vacuum c are not both equal to unity this equation reads

$$L_{GW} = \frac{G}{5c^5} \left\langle \frac{d^3}{dt^3} Q_{jk} \frac{d^3}{dt^3} Q_{jk} \right\rangle_{BHI}. \tag{4.155}$$

We note in passing that the combination of c and G appearing here gives the *luminosity* as

$$\frac{c^5}{G} = 4 \times 10^{66}\,\mathrm{Js}^{-1}. \tag{4.156}$$

Following Peters and Mathews (1963) [see also Peters (1964)] consider two point-like masses m_1 and m_2 constituting a binary system in an elliptic orbit with semi-major axis a about each other. Let a_1 and a_2 denote their respective distances from the center of mass. Then

$$m_1\,a_1 = m_2\,a_2 = \mu\,a, \tag{4.157}$$

where μ is the reduced mass defined as usual to be

$$\mu = \frac{m_1 m_2}{m_1 + m_2}. \tag{4.158}$$

Choose the z-axis as the rotation axis and denote the azimuthal angle measured from the x-axis as ψ. Then the reduced quadrupole moments at the Newtonian level become

$$Q_{xx} = \left(m_1 a_1^2 + m_2 a_2^2\right)\cos^2\psi + \text{constant} = \frac{1}{2}\mu a^2 \cos 2\psi + \text{constant}, \tag{4.159}$$

$$Q_{yy} = -\left(m_1 a_1^2 + m_2 a_2^2\right)\cos^2\psi + \text{constant} = -\frac{1}{2}\mu a^2 \cos 2\psi + \text{constant}, \tag{4.160}$$

$$Q_{xy} = \left(m_1 a_1^2 + m_2 a_2^2\right)\sin\psi\cos\psi = \frac{1}{2}\mu a^2 \sin 2\psi + \text{constant}. \tag{4.161}$$

Using the constant areal velocity law (Kepler's second law) we have

$$\frac{1}{2} r^2 \dot{\psi} = \frac{\pi ab}{T} = \frac{1}{2}\sqrt{GMa(1-e^2)}, \tag{4.162}$$

where $b = a\sqrt{1-e^2}$. We thus have

$$\dot{\psi} = \frac{\sqrt{GMa(1-e^2)}}{r^2}. \tag{4.163}$$

Now making frequent use of (4.163) we obtain the third time derivatives of the reduced quadrupole moments as

$$\frac{d^3 Q_{xx}}{dt^3} = \beta(1 + e\cos\psi)^2(2\sin 2\psi + 3e\sin\psi\cos^2\psi), \tag{4.164}$$

$$\frac{d^3 Q_{yy}}{dt^3} = -\beta(1 + e\cos\psi)^2[2\sin 2\psi + e\sin\psi(1 + 3\cos^2\psi)], \tag{4.165}$$

$$\frac{d^3 Q_{xy}}{dt^3} = -\beta(1 + e\cos\psi)^2[2\cos 2\psi - e\cos\psi(1 - 3\cos^2\psi)], \tag{4.166}$$

where we define the coefficient β by

$$\beta^2 = \frac{4G^3 m_1^2 m_2^2 M}{a^5(1-e^2)^5}. \tag{4.167}$$

Substituting these into L_{GW} we arrive at

$$L_{GW} = \frac{32}{5}\frac{G^4}{c^5}\frac{M^3\mu^2}{a^5(1-e^2)^{7/2}}\left(1+\frac{73}{24}e^2+\frac{37}{96}e^4\right). \tag{4.168}$$

Next we consider the angular momentum carried by gravitational waves. In similar fashion to the derivation of the radiated energy we start with the quadrupole formula for the radiated angular momentum,

$$\left(\frac{dL^i}{dt}\right)_{GW} = -\frac{2}{5}\epsilon^{ikl}\left\langle \frac{d^2}{dt^2}Q_{km}\frac{d^3}{dt^3}Q_{ml}\right\rangle_{BHI}. \tag{4.169}$$

Now the second time derivatives of the reduced quadrupole moments are

$$\frac{d^2Q_{xx}}{dt^2} = -\alpha(\cos 2\psi + e\cos^3\psi), \tag{4.170}$$

$$\frac{d^2Q_{yy}}{dt^2} = \alpha(\cos 2\psi + e\cos^3\psi + e\cos\psi), \tag{4.171}$$

$$\frac{d^2Q_{xy}}{dt^2} = -\alpha[\sin 2\psi + e(\sin\psi + \sin\psi\cos^2\psi)], \tag{4.172}$$

where we define the coefficient α as

$$\alpha = \beta\frac{[a(1-e^2)]^{3/2}}{(GM)^{1/2}}. \tag{4.173}$$

In particular we obtain from (4.169) the formula

$$\left(\frac{dL^z}{dt}\right)_{GW} = \frac{32}{5}\frac{G^3}{c^5}\frac{M^{5/2}\mu^2}{a^{7/2}(1-e^2)^2}\left(1+\frac{7}{8}e^2\right). \tag{4.174}$$

For a Keplerian orbit in the Newtonian approximation we have

$$a = -\frac{GM\mu}{2E}, \tag{4.175}$$

$$(L^z)^2 = M\mu^2 a(1-e^2), \tag{4.176}$$

which correspond to the total energy and angular momentum conservation laws respectively. The balance arguments yield

$$-\frac{dE}{dt} = L_{GW}, \tag{4.177}$$

$$-\frac{dL^z}{dt} = \left(\frac{dL^z}{dt}\right)_{GW}, \tag{4.178}$$

where the left-hand sides are the rates of decrease of the total energy and of the angular momentum of the source and the right-hand sides correspond to the luminosity and angular momentum carried away by gravitational waves.

By substituting (4.168) and (4.175) into (4.177) we recover the important formula (4.142) describing the shrinking rate of the orbit. Using Kepler's third law we can rewrite (4.142) as giving the decay rate of the orbital period T in the form

$$\frac{dT}{dt} = -\frac{192\pi}{5}\left(\frac{T}{2\pi}\right)^{-5/3}\frac{M^{2/3}\mu}{(1-e^2)^{7/2}}\left(1+\frac{73}{24}e^2+\frac{37}{96}e^4\right). \tag{4.179}$$

We also obtain from (4.178) the equation

$$\frac{de}{dt} = -\frac{304}{15}\frac{M^2\mu\, e}{a^4(1-e^2)^{5/2}}\left(1+\frac{121}{304}e^2\right), \tag{4.180}$$

which gives the circularization rate due to the gravitational radiation reaction in agreement with (4.121). Blanchet and Schäfer (1989) calculated the first post-Newtonian corrections to both L_{GW} and $(dL^z/dt)_{GW}$. In particular they found the 1PN corrections to the orbital period decay,

$$\begin{aligned}\left(\frac{dT}{dt}\right)_{GW} &= -\frac{192\pi}{5}\left(\frac{T}{2\pi}\right)^{-5/3}\frac{M^{2/3}\mu}{(1-e^2)^{7/2}}\\ &\quad\times\frac{(Mn)^{2/3}}{336(1-e^2)^2}\left[1273+\frac{16495}{2}e^2+\frac{42231}{8}e^4+\frac{3947}{16}e^6\right.\\ &\quad-\left(924+3381e^2+\frac{1659}{4}e^4-\frac{259}{4}e^6\right)\nu\\ &\quad\left.+\left(3297e^2+4221e^4+\frac{2331}{8}e^6\right)\frac{\delta m}{M}\right],\end{aligned} \tag{4.181}$$

where δm and ν are defined by

$$\delta m \equiv m_1 - m_2, \tag{4.182}$$

and

$$\left(\frac{\delta m}{M}\right)^2 = 1-4\nu. \tag{4.183}$$

4.9 Spin effects

Spin is a fundamental property of most astrophysical bodies as well as of elementary particles. Observation of its gravitational interaction is an important challenge. Spin interactions may take different forms. For instance it is expected that the spin of a compact rotating body in a binary system with another compact companion will gravitationally couple with the orbital angular momentum (relativistic spin–orbit coupling) and also with the spin of the companion (relativistic spin–spin coupling).

The force per mass due to the spin–orbit coupling is given by

$$B_2^i = \frac{GM}{r^3}\left(-\frac{1}{2}[\boldsymbol{S} - 3(\boldsymbol{N}\cdot\boldsymbol{S})\boldsymbol{S}\times\boldsymbol{v}]^i - \frac{3}{r}(\boldsymbol{N}\cdot\boldsymbol{S}_2)S_1^i - \frac{3}{r}(\boldsymbol{N}\cdot\boldsymbol{S}_1)S_2^i \right.$$
$$\left. -\frac{3}{r^2}[\boldsymbol{S}_1\cdot\boldsymbol{S}_2 - 5(\boldsymbol{N}\cdot\boldsymbol{S}_1)(\boldsymbol{N}\cdot\boldsymbol{S}_2)]x^i\right), \tag{4.184}$$

where $\boldsymbol{S}_1$ and $\boldsymbol{S}_2$ are the specific angular momenta, or spin vectors, of each body and we define $\boldsymbol{S}$ by

$$\boldsymbol{S} = \left(3 + \frac{m_1}{M}\right)\boldsymbol{S}_1 + \left(3 + \frac{m_2}{M}\right)\boldsymbol{S}_2. \tag{4.185}$$

Substituting $F^i = B_2^i$ into (4.120) and (4.121) we can write both of these equations in the simple vectorial forms

$$\frac{d\boldsymbol{c}}{dt} = \boldsymbol{\Omega}\times\boldsymbol{c}, \tag{4.186}$$

and

$$\frac{d\boldsymbol{e}}{dt} = \boldsymbol{\Omega}\times\boldsymbol{e}, \tag{4.187}$$

where $\boldsymbol{\Omega}$ is given by

$$\boldsymbol{\Omega} = GM\left(\frac{3n}{a(1-e^2)}\boldsymbol{k} + \frac{1}{2a^3(1-e^2)^{3/2}}[\boldsymbol{S} - 3(\boldsymbol{S}\cdot\boldsymbol{k})\boldsymbol{k}]\right.$$
$$-\frac{3}{2na^5(1-e^2)^2}\left[(\boldsymbol{S}_2\cdot\boldsymbol{k})\boldsymbol{S}_1 + (\boldsymbol{S}_1\cdot\boldsymbol{k})\boldsymbol{S}_2\right.$$
$$\left.\left. +\{\boldsymbol{S}_1\cdot\boldsymbol{S}_2 - 5(\boldsymbol{S}_1\cdot\boldsymbol{k})(\boldsymbol{S}_2\cdot\boldsymbol{k})\}\boldsymbol{k}\right]\right). \tag{4.188}$$

This set of vectorial equations can be projected along $\boldsymbol{k}$, $\boldsymbol{l}$, and $\boldsymbol{m}$ or along $\boldsymbol{P}$ and $\boldsymbol{Q}$. After lengthy but straightforward calculations we find that

$$\frac{da}{dt} = 0, \tag{4.189}$$

and

$$\frac{de}{dt} = 0, \tag{4.190}$$

which means that the spin effect at the post-Newtonian order changes neither the orbital size nor the eccentricity. We also find that

$$\frac{di}{dt} = GM\left(\frac{\boldsymbol{S}\cdot\boldsymbol{l}}{2a^3(1-e^2)^{3/2}} - \frac{3\{(\boldsymbol{S}_1\cdot\boldsymbol{k})(\boldsymbol{S}_2\cdot\boldsymbol{l}) + (\boldsymbol{S}_2\cdot\boldsymbol{k})(\boldsymbol{S}_1\cdot\boldsymbol{l})\}}{2\sqrt{GM}a^{7/2}(1-e^2)^2}\right), \tag{4.191}$$

and

$$\sin i\frac{d\Omega}{dt} = GM\left(\frac{\boldsymbol{S}\cdot\boldsymbol{m}}{2a^3(1-e^2)^{3/2}} - \frac{3\{(\boldsymbol{S}_1\cdot\boldsymbol{k})(\boldsymbol{S}_2\cdot\boldsymbol{m}) + (\boldsymbol{S}_2\cdot\boldsymbol{k})(\boldsymbol{S}_1\cdot\boldsymbol{m})\}}{2\sqrt{GM}a^{7/2}(1-e^2)^2}\right). \tag{4.192}$$

Finally if we choose the equatorial plane of the body with mass m_2 as the reference plane then we have $S_2^1 = S_2^2 = 0$ and $S_3 = |\boldsymbol{S}|$, where the lower index refers to each body and the upper index denotes the vector components, and

$$\frac{d\omega}{dt} = -\cos i \frac{d\Omega}{dt} + GM\left(\frac{3\sqrt{GM}}{a^{5/2}(1-e^2)} - \frac{\boldsymbol{S}\cdot\boldsymbol{k}}{a^3(1-e^2)^{3/2}} - \frac{3}{2\sqrt{GM}a^{7/2}(1-e^2)^2}[(\boldsymbol{S}_1\cdot\boldsymbol{S}_2) - 3(\boldsymbol{S}_1\cdot\boldsymbol{k})(\boldsymbol{S}_2\cdot\boldsymbol{k})]\right), \tag{4.193}$$

$$\frac{d\sigma}{dt} = GM\left[\left(-15 + 9\frac{m_1 m_2}{M^2}\right)\frac{\sqrt{GM}}{a^{5/2}(1-e^2)^{1/2}} + \left(6 - 7\frac{m_1 m_2}{M^2}\right)\frac{\sqrt{GM}}{a^{5/2}} - \frac{3}{2\sqrt{GM}a^{7/2}(1-e^2)^{3/2}}[(\boldsymbol{S}_1\cdot\boldsymbol{S}_2) - 3(\boldsymbol{S}_1\cdot\boldsymbol{k})(\boldsymbol{S}_2\cdot\boldsymbol{k})]\right]. \tag{4.194}$$

We have concentrated here on the effect of spin on orbital motion. The evolution of spin in gravitational fields, introduced by considering a small body specified by its mass and spin moving in a given gravitational field, has been extensively studied by Papapetrou (1951*b*) and Dixon (1970*a*, 1970*b*, 1973, 1974). The spin is found to be propagated along the world-line of the body by Fermi–Walker transport:

$$u^\mu \nabla_\mu S^\alpha = S^\mu a_\mu u^\alpha, \tag{4.195}$$

where u^μ and a_μ are the 4-velocity and 4-acceleration of the body respectively and S^μ is its spin 4-vector. If the body has a negligible mass, such as an artificial satellite, then its world-line is a geodesic for which $a_\mu = 0$. In this case the spin 4-vector is parallel transported along the world-line of the body. The spin effect has been recently observed for a newly discovered double pulsar PSR J0737-3039A/B by Breton *et al.* (2008). Their observed relativistic spin precession rate of 4.77 degrees per year is consistent with general relativity.

5
Small black holes: geometrical preliminaries

To model a small black hole moving in an external field we follow the approach of Futamase, Hogan, and Itoh (2008). The space-time model of the external field provides us with a *background* space-time. If the external field is a vacuum gravitational field, due to a large black hole for example, then the background space-time is a non-flat solution of Einstein's vacuum field equations. If we wish to model a small charged black hole moving in external gravitational and electromagnetic fields then the background space-time is a solution of the vacuum Einstein–Maxwell field equations. We include the charged case because it provides a relatively easily manageable example to illustrate the method of derivation of the equations of motion including radiation reaction. The world-line or history of the black hole will be a non-singular time-like world-line in the background space-time. The field of the small black hole will be introduced as a small perturbation of the background space-time which *inter alia* is singular on this world-line. The equations of motion of the black hole will be differential equations for this world-line, obtained using the field equations which determine the small perturbation. The perturbed space-time will be an approximate solution of the vacuum Einstein or Einstein–Maxwell field equations as appropriate. To lay the foundations for this approach to equations of motion we must study the geometry of the background space-time.

5.1 The Fermi property of the background

In the background space-time we select an arbitrary time-like world-line $r = 0$ (with r, a distance from the world-line, to be defined later) which will ultimately be the history of the small black hole when viewed in this space-time. Following the classical work of Fermi (1922) [see also O'Raifeartaigh (1958, 1962)] we can, without loss of generality, take the metric tensor in the neighbourhood of $r = 0$ to be the Minkowskian metric tensor neglecting terms of order r^2. Thus the metric tensor components $(g_{\mu\nu})$ of the background space-time satisfy

$$g_{\mu\nu} = \eta_{\mu\nu} + O(r^2), \tag{5.1}$$

with $\eta_{\mu\nu} = \text{diag}(1, -1, -1, -1)$, if we are using rectangular Cartesian coordinates and time in which to express the Minkowskian metric tensor components $\eta_{\mu\nu}$. We will denote such coordinates by (X^μ) with Greek indices taking values $0, 1, 2, 3$ with X^0 the time coordinate. We find it more convenient to use, in this chapter and in Chapter 6, a metric signature of -2 (in much the same way that a metric signature of -2 is slightly more convenient for discussing spinors in general relativity than a metric signature of $+2$). We refer to (5.1) as *the Fermi property* of the background since it will be seen to play a central part in the theory we develop in the remainder of this book. In the coordinates (X^μ) we write the parametric equations of the world-line $r = 0$ as $X^\mu = x^\mu(u)$ with u as proper time or arc length along $r = 0$. Thus if $v^\mu(u) = dx^\mu/du$ then $\eta_{\mu\nu}\, v^\mu\, v^\nu = +1$ and v^μ are the components of the 4-velocity of the particle with world-line $r = 0$. The components of the 4-acceleration will be denoted $a^\mu = dv^\mu/du$ and satisfy $\eta_{\mu\nu}\, v^\mu\, a^\nu = 0$. Although using the coordinates (X^μ) is important it will also be useful to write the Minkowskian metric tensor in curvilinear coordinates which are intimately associated with an arbitrary time-like world-line. The resulting form of the Minkowskian line-element is a member of the Robinson–Trautman (1960, 1962) family of line-elements. In this form the components of the 4-velocity and the 4-acceleration introduced above appear explicitly in the metric tensor components.

For the remainder of this section we shall consider only Minkowskian space-time, having metric tensor components $\eta_{\mu\nu}$ in the coordinates (X^μ) described above and containing the time-like world-line $X^\mu = x^\mu(u)$. Our conclusions will be incorporated, in the next section of this chapter, into the background space-time by making use of (5.1). Every event (X^μ) of Minkowskian space-time not lying on the time-like world-line can be found on a unique generator of a future null cone with vertex $S = (x^\mu(u))$ (say) on the world-line. Thus the position vector ξ^μ of this event relative to the vertex satisfies

$$\xi^\mu = X^\mu - x^\mu(u), \qquad \xi_\mu\, \xi^\mu = 0, \tag{5.2}$$

where here and in future we shall lower and raise Greek indices using the Minkowskian metric tensor and its inverse in coordinates (X^μ). Let $v^\mu(u)$ be the components of the unit tangent to the world-line at $(x^\mu(u))$. Consider the space-like geodesic passing through (X^μ) and intersecting the tangent $v^\mu(u)$ orthogonally with respect to the Minkowskian metric at some event Q. Let r be the length of the segment of this geodesic joining Q to (X^μ). This length is easily seen to be given by

$$r = v_\mu\, (X^\mu - x^\mu(u)) \geq 0. \tag{5.3}$$

This quantity is in general positive since it is the scalar product of a future pointing time-like vector and a future pointing null vector. On account of the second part of (5.2) it easily follows that this scalar product vanishes if, and only if, (X^μ) coincides with $(x^\mu(u))$. Thus r in (5.3) can be used as a measure of the distance of (X^μ) from the world-line. It is the *retarded distance* of the event (X^μ) from the world-line [see the elegant paper by Synge (1970*b*) whose line of argument we follow]. On account of the uniqueness of the null geodesic joining $S = (x^\mu(u))$ to (X^μ) we can extend $u(S)$, $v^\mu(S)$, and $a^\mu(S)$ to a scalar field and two vector fields on Minkowskian space-time by defining

$$u\,(X^\mu) = u(S), \;\; v^\mu\,(X^\mu) = v^\mu(S), \;\; a^\mu\,(X^\mu) = a^\mu(S), \tag{5.4}$$

where the latter two equations constitute parallel transport of the vectors along the null geodesic. This construction has the additional effect of making r in (5.3) a scalar field on Minkowskian space-time. To calculate derivatives of these quantities with respect to X^μ, denoted by a comma, we arbitrarily vary X^μ. Then the second of (5.2) results in

$$\xi_\mu\,(\delta X^\mu - \delta x^\mu) = 0. \tag{5.5}$$

But

$$\delta x^\mu = v^\mu\,\delta u, \tag{5.6}$$

and so we conclude from (5.5) that

$$u_{,\,\mu} = r^{-1}\xi_\mu. \tag{5.7}$$

Now it is straightforward to find further partial derivatives. Writing for convenience $k^\mu = \xi^\mu/r$, and thus $k^\mu\,k_\mu = 0$ by (5.2) and $k^\mu\,v_\mu = +1$ by (5.3), we find for example that

$$v^\mu{}_{,\,\nu} = a^\mu\,k_\nu, \qquad a^\mu{}_{,\,\nu} = \dot{a}^\mu\,k_\nu, \tag{5.8}$$

with the dot denoting differentiation with respect to u. In addition

$$\xi^\mu{}_{,\,\nu} = \delta^\mu_\nu - v^\mu\,k_\nu, \tag{5.9}$$

$$r_{,\,\mu} = v_\mu - (1 - r\,h_0)\,k_\mu, \tag{5.10}$$

and so

$$k^\mu{}_{,\,\nu} = \frac{1}{r}\,\{\delta^\mu_\nu - v^\mu\,k_\nu - k^\mu\,v_\nu + (1 - r\,h_0)\,k^\mu\,k_\nu\}, \tag{5.11}$$

with $h_0 = a_\mu\,k^\mu$. We can choose r as an affine parameter along the null geodesic integral curves of k^μ. By (5.7) we have $k_\mu = u_{,\,\mu}$ and these integral curves are the generators of the future null cones $u = \text{const.}$ [compare with the first part of (5.4)] with vertices on $r = 0$. We see from (5.11) that

$$\frac{\partial k^\mu}{\partial r} = k^\mu{}_{,\,\nu}\,k^\nu = 0, \qquad \frac{\partial k^\mu}{\partial u} = k^\mu{}_{,\,\nu}\,v^\nu = -h_0\,k^\mu. \tag{5.12}$$

To integrate the second of these equations we write

$$h_0 = P_0^{-1}\,\frac{\partial P_0}{\partial u}, \tag{5.13}$$

for some function P_0 independent of r. It then follows from (5.12) that $k^\mu = P_0^{-1}\zeta^\mu$ with ζ^μ null and independent of r and u. Thus ζ^μ can be parameterized by two real parameters x and y as, for example,

$$\zeta^\mu = \left(1 + \frac{1}{4}(x^2 + y^2), -x,\ -y,\ -1 + \frac{1}{4}(x^2 + y^2)\right), \tag{5.14}$$

with $-\infty < x < +\infty$ and $-\infty < y < +\infty$. Since $v_\mu\,k^\mu = 1$ we now have

$$P_0 = v_\mu\,\zeta^\mu = \left\{1 + \frac{1}{4}(x^2 + y^2)\right\}\,v^0 + x\,v^1 + y\,v^2 + \left\{1 - \frac{1}{4}(x^2 + y^2)\right\}\,v^3. \tag{5.15}$$

Putting $\xi^\mu = r\,k^\mu = P_0^{-1}\zeta^\mu$ into the first part of (5.2) we arrive at

$$X^\mu = x^\mu(u) + r\,P_0^{-1}\zeta^\mu. \tag{5.16}$$

This can be regarded as a coordinate transformation relating the rectangular Cartesian coordinates and time (X^μ) and curvilinear coordinates (u, r, x, y) [see for example Newman and Unti (1962)]. The line-element of Minkowskian space-time is transformed into

$$\begin{aligned} ds^2 &= \eta_{\mu\nu}\,dX^\mu\,dX^\nu, \\ &= -r^2 P_0^{-2}(dx^2 + dy^2) + 2\,du\,dr + (1 - 2\,h_0 r)\,du^2, \end{aligned} \tag{5.17}$$

which is of Robinson–Trautman form. An alternative Robinson–Trautman form of the Minkowskian line-element can be found in Molenda (1984) and in Hogan and Ellis (1989).

Differentiating (5.16) with respect to X^ν and using (5.7) and (5.10) results in

$$\delta^\mu_\nu = v^\mu\,k_\nu + k^\mu\,v_\nu - k^\mu\,k_\nu + r\,\frac{\partial k^\mu}{\partial x}\,x_{,\nu} + r\,\frac{\partial k^\mu}{\partial y}\,y_{,\nu}. \tag{5.18}$$

Multiplying this by $\partial k_\mu/\partial x$ and by $\partial k_\mu/\partial y$ yields

$$x_{,\nu} = -\frac{1}{r}P_0^2\frac{\partial k_\nu}{\partial x}, \qquad y_{,\nu} = -\frac{1}{r}P_0^2\frac{\partial k_\nu}{\partial y}. \tag{5.19}$$

When these are substituted back into (5.18) a very useful form for the metric tensor of Minkowskian space-time in the coordinates (X^μ) results, namely,

$$\eta^{\mu\nu} = -P_0^2\left(\frac{\partial k^\mu}{\partial x}\,\frac{\partial k^\nu}{\partial x} + \frac{\partial k^\mu}{\partial y}\,\frac{\partial k^\nu}{\partial y}\right) + k^\mu\,v^\nu + k^\nu\,v^\mu - k^\mu\,k^\nu. \tag{5.20}$$

A lot of use is made of this relation below.

The form of the line-element (5.17) suggests the introduction of the basis 1-forms (defining a half-null tetrad)

$$\begin{aligned} &\vartheta^1 = r\,P_0^{-1}dx, \qquad \vartheta^2 = r\,P_0^{-1}dy\,, \\ &\vartheta^3 = dr + \frac{1}{2}(1 - 2\,h_0 r)\,du, \qquad \vartheta^4 = du. \end{aligned} \tag{5.21}$$

Using (5.7), (5.10), and (5.19) these can be written in the coordinates (X^μ) as

$$\vartheta^1 = -P_0\frac{\partial k_\mu}{\partial x}\,dX^\mu = -\vartheta_1, \tag{5.22}$$

$$\vartheta^2 = -P_0\frac{\partial k_\mu}{\partial y}\,dX^\mu = -\vartheta_2, \tag{5.23}$$

$$\vartheta^3 = (v_\mu - \frac{1}{2}k_\mu\,)\,dX^\mu = \vartheta_4, \tag{5.24}$$

$$\vartheta^4 = k_\mu\,dX^\mu = \vartheta_3. \tag{5.25}$$

These will prove very useful in the sequel. In addition with $k^\mu = P_0^{-1}\zeta^\mu$ and ζ^μ given by (5.14) we have explicitly

$$-P_0\,k^\mu = -\left\{1+\frac{1}{4}(x^2+y^2)\right\}\,\delta_0^\mu + x\,\delta_1^\mu + y\,\delta_2^\mu + \left\{1-\frac{1}{4}(x^2+y^2)\right\}\,\delta_3^\mu, \qquad (5.26)$$

with $P_0(x,y,u)$ given by (5.15). Direct calculation yields the second partial derivatives of k^μ with respect to x and y which are helpful later,

$$\frac{\partial^2 k^\mu}{\partial x^2} = P_0^{-2}(v^\mu - k^\mu) - \frac{\partial}{\partial x}(\log P_0)\,\frac{\partial k^\mu}{\partial x} + \frac{\partial}{\partial y}(\log P_0)\,\frac{\partial k^\mu}{\partial y}, \qquad (5.27)$$

$$\frac{\partial^2 k^\mu}{\partial y^2} = P_0^{-2}(v^\mu - k^\mu) + \frac{\partial}{\partial x}(\log P_0)\,\frac{\partial k^\mu}{\partial x} - \frac{\partial}{\partial y}(\log P_0)\,\frac{\partial k^\mu}{\partial y}, \qquad (5.28)$$

$$\frac{\partial^2 k^\mu}{\partial x\partial y} = -\frac{\partial}{\partial y}(\log P_0)\,\frac{\partial k^\mu}{\partial x} - \frac{\partial}{\partial x}(\log P_0)\,\frac{\partial k^\mu}{\partial y}. \qquad (5.29)$$

These equations can be obtained from (5.20) and the fact that $\partial k^\mu/\partial x, \partial k^\mu/\partial y$ together with k^μ, v^μ are linearly independent (remembering also that $k^\mu\,k_\mu = 0$ and $k^\mu\,v_\mu = 1$). As an illustration of this we begin by multiplying (5.20) by $\partial k_\mu/\partial x$ which leads to

$$\frac{\partial k^\nu}{\partial x} = -P_0^2\,\frac{\partial k_\mu}{\partial x}\,\frac{\partial k^\mu}{\partial x}\,\frac{\partial k^\nu}{\partial x} - P_0^2\,\frac{\partial k_\mu}{\partial x}\,\frac{\partial k^\mu}{\partial y}\,\frac{\partial k^\nu}{\partial y}. \qquad (5.30)$$

Thus the linear independence of $\partial k^\nu/\partial x$ and $\partial k^\nu/\partial y$ gives

$$\frac{\partial k_\mu}{\partial x}\,\frac{\partial k^\mu}{\partial x} = -P_0^{-2}, \qquad \frac{\partial k_\mu}{\partial x}\,\frac{\partial k^\mu}{\partial y} = 0. \qquad (5.31)$$

Similarly multiplying (5.20) by $\partial k_\mu/\partial y$ gives

$$\frac{\partial k_\mu}{\partial y}\,\frac{\partial k^\mu}{\partial y} = -P_0^{-2}. \qquad (5.32)$$

Now differentiating (5.20) with respect to x and then multiplying by $\partial k_\nu/\partial x$ and using (5.31) results in

$$\frac{\partial^2 k^\mu}{\partial x^2} = P_0^{-2}(v^\mu - k^\mu) - 2\,P_0^{-1}\frac{\partial P_0}{\partial x}\,\frac{\partial k^\mu}{\partial x} + P_0^2\,\frac{\partial k_\nu}{\partial x}\,\frac{\partial^2 k^\nu}{\partial x^2}\,\frac{\partial k^\mu}{\partial x} + P_0^2\frac{\partial k_\nu}{\partial x}\,\frac{\partial^2 k^\nu}{\partial x\partial y}\,\frac{\partial k^\mu}{\partial y}. \qquad (5.33)$$

Differentiating the first of (5.31) with respect to x gives

$$\frac{\partial k_\nu}{\partial x}\,\frac{\partial^2 k^\nu}{\partial x^2} = P_0^{-3}\frac{\partial P_0}{\partial x}, \qquad (5.34)$$

while differentiating the first of (5.31) with respect to y gives

$$\frac{\partial k_\nu}{\partial x}\,\frac{\partial^2 k^\nu}{\partial y\partial x} = P_0^{-3}\frac{\partial P_0}{\partial y}, \qquad (5.35)$$

and using these in (5.33) results in (5.27). The equations (5.28) and (5.29) are similarly derived from (5.20).

5.2 The line-element of the background

Returning to the Fermi property (5.1) we see that the Minkowskian metric tensor (η_{ij}) can be written in the form given via the line-element (5.17) in coordinates (x, y, r, u) using the transformation (5.16) and, in effect, neglecting $O(r^2)$–terms. The null hypersurfaces $u = \text{const.}$ are, near $r = 0$, future null cones with vertices on $r = 0$. The coordinate u is arc length along $r = 0$. The coordinate r is an affine parameter along the generators of the null hypersurfaces and the generators are each specified by a choice of the pair of coordinates (x, y). All of this holds *near* $r = 0$. It would be natural and logical to extend this construction at least some distance outside this tubular neighborhood of the time-like world-line $r = 0$. This can be done in the following way. We begin by constructing a completely general line-element for the background space-time based on a family of null hypersurfaces (Hogan and Trautman, 1987) [a derivation is given in Futamase and Hogan (1993)]. Let $\{z^\mu\}$ be a local coordinate system in the background space-time in terms of which (k^μ) are the components of a null vector field, so that

$$g_{\mu\nu}\, k^\mu\, k^\nu = 0, \tag{5.36}$$

where (g_{ij}) are the components of the metric tensor of the background space-time in coordinates $\{z^\mu\}$. In addition we take

$$k_\mu\, dz^\mu = du, \tag{5.37}$$

so that k^μ is tangent to the null geodesic generators of the hypersurfaces $u(z^\mu) = \text{const.}$ Let $z^\mu = z^\mu(r)$ be the equation of a typical generator, with r an affine parameter, so that

$$k^\mu\, \frac{\partial}{\partial z^\mu} = \frac{\partial}{\partial r}. \tag{5.38}$$

Choosing as coordinates (u, r, z^2, z^3) the line-element can be written in the form

$$ds^2 = -q_{AB}\,(dz^A + a^A\, du)\,(dz^B + a^B\, du) + 2\, du\, dr + c\, du^2, \tag{5.39}$$

with $q_{AB} = q_{BA}, a^A, c$ functions of (u, r, z^A) and capital Latin indices take values 2, 3. We can factor the determinant out of the 2×2 symmetric matrix (q_{AB}) by writing

$$q_{AB} = P^2\, \gamma_{AB}, \tag{5.40}$$

and choosing the function P so that the determinant of the 2×2 matrix (γ_{AB}) is unity. We write, for later convenience,

$$P = r\, p^{-1}, \tag{5.41}$$

where p is a function of (u, r, z^A) and we parameterize (γ_{AB}) by

$$(\gamma_{AB}) = \begin{pmatrix} e^{2\,\alpha}\, \cosh 2\beta & \sinh 2\beta \\ \sinh 2\beta & e^{-2\,\alpha}\, \cosh 2\beta \end{pmatrix}, \tag{5.42}$$

where α, β are functions of (u, r, z^A). Finally writing $z^2 = x$ and $z^3 = y$ and putting

$$a = a^1\, e^{\alpha}\, \cosh\beta + a^2\, e^{-\alpha}\, \sinh\beta, \tag{5.43}$$

$$b = a^1\, e^{\alpha}\, \sinh\beta + a^2\, e^{-\alpha}\, \cosh\beta, \tag{5.44}$$

the line-element (5.39), in terms of a convenient basis of 1-forms, reads

$$ds^2 = -(\vartheta^1)^2 - (\vartheta^2)^2 + 2\,\vartheta^3\,\vartheta^4, \tag{5.45}$$

with

$$\vartheta^1 = r\,p^{-1}(e^{\alpha}\, \cosh\beta\, dx + e^{-\alpha}\, \sinh\beta\, dy + a\, du), \tag{5.46}$$

$$\vartheta^2 = r\,p^{-1}(e^{\alpha}\, \sinh\beta\, dx + e^{-\alpha}\, \cosh\beta\, dy + b\, du), \tag{5.47}$$

$$\vartheta^3 = dr + \frac{c}{2}\, du, \tag{5.48}$$

$$\vartheta^4 = du. \tag{5.49}$$

This line-element is completely general, containing six functions $p, \alpha, \beta, a, b, c$ of the four coordinates u, r, x, y. It is therefore equivalent to line-elements constructed by Sachs (1962) and by Newman and Unti (1962). The particular form given here was designed in Hogan and Trautman (1987) to allow the Robinson–Trautman line-elements to emerge as a convenient special case (obtained by putting $\alpha = \beta = 0$). It is this property of the line-element that makes it useful in the current context.

The important properties of the line-element given by (5.45)–(5.49) which we emphasize are: (a) the hypersurfaces $u =$ const. are null and are generated by the null geodesic integral curves of the vector field $\partial/\partial r$, (b) the coordinate r is an affine parameter along these null geodesics, (c) the coordinates x, y label each such curve in a null hypersurface $u =$ const., and (d) the null geodesic congruence has complex shear (see Appendix C)

$$\sigma = \frac{\partial\alpha}{\partial r}\, \cosh 2\beta + i\, \frac{\partial\beta}{\partial r}, \tag{5.50}$$

and real expansion

$$\theta = \frac{\partial}{\partial r}\, \log(r\,p^{-1}). \tag{5.51}$$

In view of the Minkowskian form (5.17) we can now re-express the Fermi property (5.1) for small r using the line-element (5.45)–(5.49) by assuming that the six functions appearing in the line-element have the following expansions in positive powers of r:

$$p = P_0(1 + q_2\, r^2 + q_3\, r^3 + \dots\,), \tag{5.52}$$

$$\alpha = \alpha_2\, r^2 + \alpha_3\, r^3 \dots, \tag{5.53}$$

$$\beta = \beta_2\, r^2 + \beta_3\, r^3 \dots, \tag{5.54}$$

$$a = a_1\, r + a_2\, r^2 \dots, \tag{5.55}$$

$$b = b_1\, r + b_2\, r^2 \ldots, \tag{5.56}$$

$$c = c_0 + c_1\, r + c_2\, r^2 + \ldots. \tag{5.57}$$

The coefficients of the various powers of r in these expansions are functions of x, y, u. The starting terms in the expansions are chosen to be consistent with (5.1) and (5.17) and thus P_0 in (5.52) is given by (5.15) while

$$c_0 = 1 = \Delta \log P_0, \qquad \Delta = P_0^2 \left(\frac{\partial^2}{\partial x^2} + \frac{\partial^2}{\partial y^2}\right), \tag{5.58}$$

$$c_1 = -2\, h_0, \tag{5.59}$$

with

$$h_0 = \eta_{\mu\nu}\, a^\mu\, k^\nu = \frac{\partial}{\partial u}(\log P_0), \tag{5.60}$$

as introduced in (5.10) and (5.13). The operator Δ in (5.58) is the Laplacian on the unit 2-sphere. Now the complex shear and real expansion of the integral curves of $\partial/\partial r$ are given respectively by (see Appendix C)

$$\sigma = 2\,(\alpha_2 + i\beta_2)\, r + 3\,(\alpha_3 + i\beta_3)\, r^2 + \ldots, \tag{5.61}$$

$$\theta = \frac{1}{r} - 2\, q_2\, r - 3\, q_3\, r^2 + \ldots. \tag{5.62}$$

Hence near $r = 0$ (i.e. for small values of r) the null hypersurfaces $u =$ const. resemble future null cones with vertices on the world-line $r = 0$.

5.3 The field equations near $r = 0$

In order to apply the construction here to a small charged black hole moving in external gravitational and electromagnetic fields, we will consider a background space-time which is a solution of the Einstein–Maxwell vacuum field equations. The pure gravitational case can then be obtained by removing the background Maxwell field. The background Maxwell field is derived from a potential 1-form which, in the coordinates u, r, x, y, is written

$$A = L\, dx + M\, dy + K\, du. \tag{5.63}$$

This must lead to a Maxwell field (the *external* Maxwell field) which is non-singular on the world-line $r = 0$. This is achieved by choosing, in the neighbourhood of the world-line, the following expansions for the coefficients of the differentials of the coordinates in (5.63):

$$L = r^2 L_2 + r^3 L_3 + \ldots, \tag{5.64}$$

$$M = r^2 M_2 + r^3 M_3 + \ldots, \tag{5.65}$$

$$K = r\, K_1 + r^2 K_2 + \ldots. \tag{5.66}$$

The coefficients of the powers of r here are functions of x, y, u. In (5.63) we could have included a term $+W\,dr$ and then we would have to expand W as $W = r\,W_1 + r^2 W_2 + r^3 W_3 + \ldots$, in order not to introduce a term into the Maxwell field which is singular on $r = 0$. Now this term $+W\,dr$ can be removed by a gauge transformation without affecting the forms of the expansions (5.64)–(5.66).

The electromagnetic field calculated with the 1-form (5.63) has the form

$$F = dA = \frac{1}{2} F_{ab}\,\vartheta^a \wedge \vartheta^b, \tag{5.67}$$

with $F_{ab} = -F_{ba}$ and ϑ^a given by (5.46)–(5.49) with $a, b = 1, 2, 3, 4$ tetrad indices. Imposing Maxwell's vacuum field equations $d^*F = 0$, where the star denotes the Hodge dual, we find in leading order in powers of the coordinate r the following equations to be satisfied by the functions $L_2,\ M_2,\ K_1$:

$$K_1 = P_0^2 \left(\frac{\partial L_2}{\partial x} + \frac{\partial M_2}{\partial y}\right), \qquad \Delta K_1 + 2\,P_0^2 \left(\frac{\partial L_2}{\partial x} + \frac{\partial M_2}{\partial y}\right) = 0, \tag{5.68}$$

and

$$\frac{\partial K_1}{\partial x} + 2\,L_2 - \frac{\partial}{\partial y}\left\{P_0^2 \left(\frac{\partial M_2}{\partial x} - \frac{\partial L_2}{\partial y}\right)\right\} = 0, \tag{5.69}$$

$$\frac{\partial K_1}{\partial y} + 2\,M_2 + \frac{\partial}{\partial x}\left\{P_0^2 \left(\frac{\partial M_2}{\partial x} - \frac{\partial L_2}{\partial y}\right)\right\} = 0, \tag{5.70}$$

with Δ the operator introduced in (5.58). We see that the second part of (5.68) is a consequence of (5.69) and (5.70). Also (5.68) implies that K_1 is an $l = 1$ spherical harmonic,

$$\Delta K_1 + 2\,K_1 = 0. \tag{5.71}$$

In this context a spherical harmonic of order l is the smooth solution $\mathcal{F}$ of $\Delta\mathcal{F} + l(l+1)\,\mathcal{F} = 0$. For our purposes the only functions of x, y, u appearing in (5.64)–(5.66) that we will require in the sequel are L_2, M_2, and K_1 and so we need not pursue Maxwell's vacuum field equations to higher order in powers of r.

The leading terms in the tetrad components of the Maxwell field F_{ab} that we will require are now given by

$$F_{13} = -2\,P_0\,L_2 + O(r), F_{23} = -2\,P_0\,M_2 + O(r), F_{34} = K_1 + O(r). \tag{5.72}$$

On $r = 0$ we can replace the basis 1-forms (5.46)–(5.49) by their Minkowskian counterparts (5.22)–(5.25) and the coordinates x, y, r, u by the coordinates (X^μ). Thus evaluating (5.72) on $r = 0$ yields

$$L_2 = \frac{1}{2} F_{\mu\nu}(u)\,k^\mu\,\frac{\partial k^\nu}{\partial x}, \qquad M_2 = \frac{1}{2} F_{\mu\nu}(u)\,k^\mu\,\frac{\partial k^\nu}{\partial y}, \qquad K_1 = F_{\mu\nu}\,k^\mu\,v^\nu. \tag{5.73}$$

where $F_{\mu\nu}(u) = -F_{\nu\mu}(u)$ are the components of the external Maxwell field calculated in the coordinates (X^μ) on $r = 0$. We can now verify directly that these expressions for L_2, M_2, K_1 satisfy the Maxwell equations (5.68)–(5.71) above. For example on substituting L_2, M_2 from (5.73) into the first part of (5.68) yields

$$K_1 = P_0^2\left(\frac{\partial L_2}{\partial x}+\frac{\partial M_2}{\partial y}\right) = \frac{1}{2}F_{\mu\nu}(u)k^\mu\,\Delta k^\nu = F_{\mu\nu}(u)\,k^\mu\,v^\nu, \tag{5.74}$$

with the last equality following from (5.27) and (5.28) added together to give

$$\Delta k^\mu + 2\,k^\mu = 2\,v^\mu. \tag{5.75}$$

Taking the electromagnetic field above as source, Einstein's field equations for the background space-time read

$$R_{ab} = 2\,E_{ab}, \tag{5.76}$$

where R_{ab} are the components of the Ricci tensor calculated on the tetrad given via the 1-forms (5.46)–(5.49). E_{ab} are the tetrad components of the electromagnetic energy–momentum tensor calculated using the Maxwell tensor (5.67) according to the formula

$$E_{ab} = F_{ca}\,F^c{}_b - \frac{1}{4}\,g_{ab}\,F_{cd}\,F^{cd}. \tag{5.77}$$

Here g_{ab} are the tetrad components of the metric tensor and indices on F_{ab} are raised using its inverse. We will give here only the consequences of the field equations (5.76) near $r = 0$ which will be useful later. To satisfy $R_{33} = 2\,E_{33} + O(r)$ we must have

$$q_2 = \frac{2}{3}\,P_0^2\,(L_2^2 + M_2^2), \tag{5.78}$$

for q_2 appearing in (5.52). With L_2, M_2 given by (5.73), and using (5.20) this can be written

$$q_2 = -\frac{1}{6}\,F^\lambda{}_\mu\,F_{\lambda\nu}\,k^\mu\,k^\nu. \tag{5.79}$$

Here $F_{\lambda\nu} = F_{\lambda\nu}(u)$, with $F^\lambda{}_\mu = \eta^{\lambda\sigma}F_{\sigma\mu}(u)$, is the Maxwell tensor in coordinates (X^μ) evaluated on $r = 0$. Now to have

$$R_{12} = 2\,E_{12} + O(r),\ \ R_{11} - R_{22} = 2\,(E_{11} - E_{22}) + O(r), \tag{5.80}$$

requires

$$2\,(\alpha_2 + i\beta_2) = -\frac{\partial}{\partial\bar\zeta}\left(a_1 + ib_1 + 4\,P_0^2\frac{\partial q_2}{\partial\zeta}\right), \tag{5.81}$$

where $\zeta = x + iy$ and a bar will denote complex conjugation. Next, for $A = 1, 2$,

$$R_{A3} = 2\,E_{A3} + O(r), \tag{5.82}$$

provided

$$a_1 + ib_1 + 4\,P_0^2\frac{\partial q_2}{\partial\bar\zeta} = 2\,P_0^4\frac{\partial}{\partial\zeta}(P_0^{-2}(\alpha_2 + i\beta_2)). \tag{5.83}$$

Putting (5.81) and (5.83) together we arrive at

$$\frac{\partial}{\partial\bar\zeta}\left\{P_0^4\frac{\partial}{\partial\zeta}(P_0^{-2}(\alpha_2+i\beta_2))\right\}=-(\alpha_2+i\beta_2). \tag{5.84}$$

The approximate field equations $R_{AA}=2\,E_{AA}+O(r)$ and $R_{34}=2\,E_{34}+O(r)$ both yield the same equation for the function c_2 appearing in (5.57) which will not be used in the sequel. The remaining field equations in the approximate form

$$R_{14}+iR_{24}=2\,(E_{14}+iE_{24})+O(r),\ \ R_{44}=2\,E_{44}+O(r), \tag{5.85}$$

are automatically satisfied. To check this requires some lengthy calculations. With the Einstein–Maxwell vacuum field equations satisfied in the neighbourhood of the world-line $r=0$ we find the following tetrad components C_{abcd} of the Weyl conformal curvature tensor in the neighbourhood of $r=0$:

$$C_{1313}+iC_{1323}=6\,(\alpha_2+i\beta_2)+O(r), \tag{5.86}$$

$$C_{3431}+iC_{3432}=\frac{3}{2}P_0^{-1}\left(a_1+ib_1+4\,P_0^2\frac{\partial q_2}{\partial\bar\zeta}\right)+O(r). \tag{5.87}$$

Using the 1-forms defined in (5.22)–(5.25) and q_2 given by (5.79) we deduce from these that

$$\alpha_2=\frac{1}{6}P_0^2C_{\mu\nu\rho\sigma}\,k^\mu\,\frac{\partial k^\nu}{\partial x}\,k^\rho\,\frac{\partial k^\sigma}{\partial x}, \tag{5.88}$$

$$\beta_2=\frac{1}{6}P_0^2C_{\mu\nu\rho\sigma}\,k^\mu\,\frac{\partial k^\nu}{\partial x}\,k^\rho\,\frac{\partial k^\sigma}{\partial y}, \tag{5.89}$$

$$a_1=\frac{2}{3}P_0^2\left(C_{\mu\nu\rho\sigma}\,k^\mu\,v^\nu\,k^\rho\,\frac{\partial k^\sigma}{\partial x}+F^\lambda{}_\mu\,F_{\lambda\nu}\,k^\mu\,\frac{\partial k^\nu}{\partial x}\right), \tag{5.90}$$

$$b_1=\frac{2}{3}P_0^2\left(C_{\mu\nu\rho\sigma}\,k^\mu\,v^\nu\,k^\rho\,\frac{\partial k^\sigma}{\partial y}+F^\lambda{}_\mu\,F_{\lambda\nu}\,k^\mu\,\frac{\partial k^\nu}{\partial y}\right). \tag{5.91}$$

Here $C_{\mu\nu\rho\sigma}(u)$ are the components of the Weyl tensor of this "background" space-time calculated on $r=0$ in the coordinates (X^μ). Hence they satisfy $C_{\mu\nu\rho\sigma}=-C_{\nu\mu\rho\sigma}=-C_{\mu\nu\sigma\rho}=C_{\rho\sigma\mu\nu}$ and $\eta^{\mu\sigma}\,C_{\mu\nu\rho\sigma}=0$. Among the coefficients of the powers of r in (5.52)–(5.57) we shall only require here the functions $q_2,\alpha_2,\beta_2,a_1,b_1$ given by (5.79) and (5.88)–(5.91), along with P_0, c_0, and c_1 given in (5.15), (5.58), and (5.59) respectively. Using the derivatives of k^μ listed in (5.27)–(5.29) one can directly verify that the field equations (5.81) and (5.83) (and thus (5.84)) are satisfied by (5.79) and (5.88)–(5.91). If the background Maxwell field vanishes then in particular $F_{\mu\nu}(u)=0$ in the formulas above and the Weyl tensor components $C_{\mu\nu\rho\sigma}(u)$ calculated on the world-line $r=0$ in coordinates (X^μ) are replaced by the Riemann curvature tensor components $R_{\mu\nu\rho\sigma}(u)$.

6

Small charged black holes: equations of motion

A small charged black hole moving in external gravitational and electromagnetic fields is useful for presenting our approach to deriving its equations of motion from the Einstein–Maxwell field equations as it provides a relatively accessible example illustrating how an external 4-force enters the equations of motion as well as a radiation reaction 4-force in addition to so-called tail terms exhibiting retardation effects. As such it is a template for more physically realistic applications of the method, in particular to the equations of motion of a small Schwarzschild black hole moving in an external gravitational field, which is partially included in the current illustrative example when the charge is put equal to zero.

6.1 Black hole as perturbation of background

We introduce the small charged black hole as a perturbation of the background Einstein–Maxwell space-time described in Chapter 5. With the mass m and charge e considered small of first order we write $m = O_1$ and $e = O_1$. The perturbation is introduced by modifying the expansions (5.52)–(5.57) as follows:

$$p = \hat{P}_0(1 + \hat{q}_2\, r^2 + \hat{q}_3\, r^3 \dots), \tag{6.1}$$

$$\alpha = \hat{\alpha}_2\, r^2 + \hat{\alpha}_3\, r^3 + \dots, \tag{6.2}$$

$$\beta = \hat{\beta}_2\, r^2 + \hat{\beta}_3\, r^3 + \dots, \tag{6.3}$$

$$a = \frac{\hat{a}_{-1}}{r} + \hat{a}_0 + \hat{a}_1\, r + \hat{a}_2\, r^2 + \dots, \tag{6.4}$$

$$b = \frac{\hat{b}_{-1}}{r} + \hat{b}_0 + \hat{b}_1\, r + \hat{b}_2\, r^2 + \dots, \tag{6.5}$$

$$c = \frac{e^2}{r^2} - \frac{2\,(m + 2\,\hat{f}_{-1})}{r} + \hat{c}_0 + \hat{c}_1\, r + \hat{c}_2\, r^2 \dots. \tag{6.6}$$

The coefficients of the powers of r are functions of x, y, u. The hats are introduced to distinguish these coefficients from the background coefficients of Chapter 5. Functions here that have non-zero background values are assumed to differ from their background values by O_1-terms. Functions which vanish in the background are assumed to have orders of magnitude: $\hat{a}_{-1} = O_1,\ \hat{a}_0 = O_1,\ \hat{b}_{-1} = O_1,\ \hat{b}_0 = O_1,\ \hat{f}_{-1} = O_2$.

The perturbed potential 1-form is taken to have the form (5.63) with the expansions (5.64)–(5.66) modified to read,

$$L = \hat{L}_0 + r^2\hat{L}_2 + r^3\hat{L}_3 + \dots, \tag{6.7}$$

$$M = \hat{M}_0 + r^2\hat{M}_2 + r^3\hat{M}_3 + \dots, \tag{6.8}$$

$$K = \frac{(e+\hat{K}_{-1})}{r} + \hat{K}_0 + r\,\hat{K}_1 + r^2\hat{K}_2 + \dots. \tag{6.9}$$

Again functions here which are non-zero in the background are assumed to differ from their background values by O_1-terms and those functions which vanish in the background are taken to have the orders of magnitude: $\hat{L}_0 = O_1,\ \hat{M}_0 = O_1,\ \hat{K}_{-1} = O_2,\ \hat{K}_0 = O_1$. All of these expansions have been chosen in order to satisfy the Reissner–Nordström small black hole limit, namely, as $r \to 0$ and at the same time $e \to 0$ and $m \to 0$ in such a way that the ratios e/r and m/r remain finite, then the perturbed gravitational field (Weyl tensor) is predominantly that of the Reissner–Nordström black hole and the perturbed electromagnetic field (Maxwell tensor) is predominantly the Coulomb field. A careful study suggests that these expansions are the minimal choice of expansions necessary to satisfy this small black hole limit. We consider, however, that these assumptions restrict the model that we are constructing of a small black hole moving in external electromagnetic and gravitational fields and that their generalization or otherwise is a topic for further study. In order to simplify the presentation we make the further assumption that near $r = 0$ the potential 1-form is predominantly the Liénard–Wiechert 1-form [given by $A = e\,(r^{-1} - h_0)\,du$ up to a gauge term]. This is achieved with the further specialization of the functions $\hat{L}_0, \hat{M}_0, \hat{K}_0$ to satisfy

$$\hat{L}_0 = O_2\ ,\ \hat{M}_0 = O_2\ ,\ \hat{K}_0 = -e\,h_0 + O_2, \tag{6.10}$$

with h_0 given by (5.60).

We emphasise that the approach we are describing here is *not* one of "matched asymptotic expansions" in which so-called "near zone" and "far zone" expansions of field variables are matched in an intermediate "buffer zone". In our perturbed space-time each function appearing in the metric tensor and the potential 1-form has only one expansion in powers of r given by (6.1)–(6.6) and (6.7)–(6.9) and all of our deductions, including the equations of motion, are made using these expansions and the vacuum Einstein–Maxwell field equations.

The expansions we have chosen here ensure that the null hypersurfaces $u =$ constant in the perturbed space-time, which will play the role of the histories of the wave fronts of the radiation produced by the black hole motion, are approximately future null cones for small r. Thus the wave fronts can be approximately 2-spheres near the black hole. Neglecting $O(r^4)$-terms the line-elements induced on these null hypersurfaces are given by

$$ds_0^2 = -r^2\hat{P}_0^{-2}\,(dx^2+dy^2) \qquad \text{with} \qquad \hat{P}_0 = P_0\,(1 + Q_1 + Q_2 + O_3), \tag{6.11}$$

with P_0 given by (5.15) and $Q_1 = O_1,\ Q_2 = O_2$. We shall require the perturbations of these 2-spheres, described by the functions Q_1 and Q_2 here, to be smooth so that as functions of x, y they are non-singular (in particular at $x = \pm\infty$ and/or $y = \pm\infty$) so that no "directional singularities" occur violating the notion of an isolated black hole. We will insist that in general in solving the Einstein–Maxwell vacuum field equations for the perturbed space-time and the perturbed electromagnetic field, such directional singularities are unacceptable. Then it will follow from the Einstein–Maxwell field equations that *necessary conditions for the 2-surfaces with line-elements (6.11) to be smooth, non-trivial deformations of 2-spheres will be the equations of motion of the black hole.* Such an origin of equations of motion can be found, for example, in Newman and Posadas (1969), Robinson and Robinson (1972), Hogan and Imaeda (1979), where the space-time is less general than here and in Hogan and Robinson (1985), where the external field is weak.

6.2 The vacuum Einstein–Maxwell field equations

The left-hand sides of the vacuum Einstein–Maxwell field equations for the perturbed space-time are now power series in r, each with a finite number of terms involving inverse powers of r and an infinite number of terms involving positive powers of r. The coefficients of the leading terms in each case are listed in Appendix D. They are functions of x, y, u and we require them to be small in terms of e and m with just sufficient accuracy to enable us to derive the functions in the expansions (6.1)–(6.6) and (6.7)–(6.9) required for the equations of motion of the black hole with an O_3-error, to include electromagnetic radiation reaction etc. The errors that we tolerate to achieve this are indicated in Appendix D. Some of these errors could be reduced but with no influence on the equations of motion to the accuracy that we calculate here. We begin by calculating the equations of motion with an O_2-error. In this case we require from Einstein's equations, for the moment requiring (D.33) and (D.39) only to be small of second order,

$$\hat{a}_{-1} = -4\,e\,P_0^2 L_2 + O_2 = O_1\ , \qquad \hat{b}_{-1} = -4\,e\,P_0^2 M_2 + O_2 = O_1, \tag{6.12}$$

and thus from Maxwell's equations, with (D.21) required at this stage to be small of first order, we have

$$\hat{K}_1 = P_0^2\left(\frac{\partial L_2}{\partial x} + \frac{\partial M_2}{\partial y}\right) + O_1 = K_1 + O_1, \tag{6.13}$$

with K_1 given in (5.74). We also need

$$\hat{c}_0 = 1 + \Delta Q_1 + 2\,Q_1 + 8\,e\,F_{\mu\nu}\,k^\mu\,v^\nu + O_2, \tag{6.14}$$

which emerges from the requirement that (D.25) be small of second order. When these are substituted into the final Einstein field equation, given by requiring (D.52) for the moment to be small of second order, the differential equation that emerges for Q_1 is

$$-\frac{1}{2}\Delta(\Delta Q_1 + 2\,Q_1) = 6\,m\,a_\mu\,p^\mu - 6\,e\,F_{\mu\nu}\,p^\mu\,v^\nu + O_2. \tag{6.15}$$

Here $p^\mu = h^\mu_\nu\,k^\nu$ with $h^\mu_\nu = \delta^\mu_\nu - v^\mu\,v_\nu$ is the projection of k^μ orthogonal to v^μ and satisfies $p^\mu\,p_\mu = -1, p^\mu\,v_\mu = 0$ and since $p^\mu = k^\mu - v^\mu$ we see from (5.27) and (5.28) that $\Delta p^\mu + 2\,p^\mu = 0$, so that each component of p^μ is an $l = 1$ spherical harmonic, remembering that a spherical harmonic Q of order l is a smooth solution of $\Delta Q + l(l+1)\,Q = 0$. Integrating (6.15), discarding the homogeneous solution with directional singularities, we have

$$\Delta Q_1 + 2\,Q_1 = 6\,m\,a_\mu\,p^\mu - 6\,e\,F_{\mu\nu}\,p^\mu\,v^\nu + A(u) + O_2, \tag{6.16}$$

where $A(u) = O_1$ is arbitrary. The first two terms on the right-hand side will now lead to directional singularities in Q_1 unless $ma_\mu\,p^\mu - e\,F_{\mu\nu}\,p^\mu\,v^\nu = O_2$ for all p^μ, which implies *the equations of motion in first approximation*,

$$ma_\mu = e\,F_{\mu\nu}\,v^\nu + O_2. \tag{6.17}$$

Hence we see the Lorentz 4-force due to the external electromagnetic field appearing. With (6.17) satisfied, equation (6.16) simplifies and the directional singularity free Q_1 is a linear combination of an $l = 1$ spherical harmonic, the smooth homogeneous solution, and an $l = 0$ spherical harmonic, the particular integral $A(u)/2$. In general perturbations of (6.11) described by Q_1 or Q_2 which are $l = 0$ or $l = 1$ spherical harmonics are trivial. When the 2-surfaces with line-element (6.11) are viewed against the Euclidean 3-space background such terms in Q_1 or Q_2 merely infinitesimally change the radius of the sphere (if $l = 0$) or infinitesimally displace the origin of the sphere (if $l = 1$). We shall consistently discard such "perturbations" which means that in this case we shall take $Q_1 = 0$.

From now on we take $Q_1 = 0$ in (6.11) and our objective is to solve the Einstein–Maxwell vacuum field equations for the perturbed space-time and the perturbed Maxwell field with sufficient accuracy to enable us derive the function Q_2 in (6.11) and the equations of motion with an O_3-error. The functions of x, y, u appearing in the potential 1-form, as coefficients in the expansions (6.7)–(6.9), that are involved are $\hat{L}_0$, $\hat{M}_0$, $\hat{K}_{-1}$, and the O_1-part of $\hat{K}_1$ and $\hat{K}_0$. Two of Maxwell's equations [(D.13) and (D.16) small of third order] read

$$\frac{\partial \hat{K}_{-1}}{\partial x} + 6\,e^2 L_2 + \frac{\partial}{\partial y}\left\{P_0^2\left(\frac{\partial \hat{M}_0}{\partial x} - \frac{\partial \hat{L}_0}{\partial y}\right)\right\} = O_3, \tag{6.18}$$

and

$$\frac{\partial \hat{K}_{-1}}{\partial y} + 6\,e^2 M_2 - \frac{\partial}{\partial x}\left\{P_0^2\left(\frac{\partial \hat{M}_0}{\partial x} - \frac{\partial \hat{L}_0}{\partial y}\right)\right\} = O_3. \tag{6.19}$$

These imply

$$\Delta \hat{K}_{-1} + 6\,e^2 F_{\mu\nu}\,k^\mu\,v^\nu = O_3, \tag{6.20}$$

which incidentally is (D.19), another of the Maxwell equations. This latter equation can immediately be integrated without introducing directional singularities to yield

$$\hat{K}_{-1} = 3\,e^2 F_{\mu\nu}\,k^\mu\,v^\nu + e^2 K(u) + O_3, \tag{6.21}$$

where $K(u) = O_0$ is an arbitrary function. From (6.18) and (6.19) we obtain

$$\Delta\left\{P_0^2\left(\frac{\partial \hat{M}_0}{\partial x} - \frac{\partial \hat{L}_0}{\partial y}\right)\right\} - 6\,e^2 P_0^2 F_{\mu\nu}\,\frac{\partial k^\mu}{\partial x}\,\frac{\partial k^\nu}{\partial y} = O_3. \tag{6.22}$$

Now the second term on the left here is an $l = 1$ spherical harmonic. In fact we can write

$$P_0^2 F_{\mu\nu}\,\frac{\partial k^\mu}{\partial x}\,\frac{\partial k^\nu}{\partial y} = F^*_{\mu\nu}\,k^\mu\,v^\nu, \tag{6.23}$$

where $F^*_{\mu\nu}(u)$ is the dual of $F_{\mu\nu}(u)$, on account of the relationship

$$\epsilon_{\mu\nu\rho\sigma}\,k^\mu\,\frac{\partial k^\nu}{\partial x} = k_\rho\,\frac{\partial k_\sigma}{\partial y} - k_\sigma\,\frac{\partial k_\rho}{\partial y}, \tag{6.24}$$

where $\epsilon_{\mu\nu\rho\sigma}$ is the Levi-Civita permutation symbol in four dimensions and k^μ is given by (5.26). We integrate (6.22) to obtain

$$P_0^2\left(\frac{\partial \hat{M}_0}{\partial x} - \frac{\partial \hat{L}_0}{\partial y}\right) = -3\,e^2 F^*_{\mu\nu}\,k^\mu\,v^\nu + e^2 S(u) + O_3, \tag{6.25}$$

with $S(u) = O_0$ arbitrary. This is the extent to which we shall require a knowledge of $\hat{L}_0$ and $\hat{M}_0$. Substituting (6.21) and (6.25) into the first-order equations (6.18) and (6.19) verifies that they are now satisfied.We now look for the O_1-parts of $\hat{L}_2$ and $\hat{M}_2$. We shall write $\hat{L}_2 = L_2 + l_2 + O_2$ and $\hat{M}_2 = M_2 + m_2 + O_2$ with $l_2 = O_1$ and $m_2 = O_1$. We can obtain l_2 and m_2 indirectly as follows: in the light of (6.21) and (6.25) two of Einstein's field equations yield algebraically [(D.33) and (D.39) required to be O_3]

$$\begin{aligned}\hat{a}_{-1} &= -4\,e\,P_0^2(L_2 + l_2) + 6\,e^2 P_0^2 F^\sigma{}_\mu\,F_{\sigma\nu}\,k^\mu\,\frac{\partial k^\nu}{\partial x} + 4\,e^2 P_0^2 M_2\,S(u)\\ &\quad -4\,e^2 P_0^2 L_2\,K(u) + O_3,\end{aligned} \tag{6.26}$$

$$\begin{aligned}\hat{b}_{-1} &= -4\,e\,P_0^2(M_2 + m_2) + 6\,e^2 P_0^2 F^\sigma{}_\mu\,F_{\sigma\nu}\,k^\mu\,\frac{\partial k^\nu}{\partial y} - 4\,e^2 P_0^2 L_2\,S(u)\\ &\quad -4\,e^2 P_0^2 M_2\,K(u) + O_3.\end{aligned} \tag{6.27}$$

Define two functions $\hat{A}$ and $\hat{B}$ by

$$\hat{a}_{-1} + 8\,P_0^2 L_2 \hat{K}_{-1} = -4\,e\,P_0^2 L_2 + \hat{A}, \tag{6.28}$$

and

$$\hat{b}_{-1} + 8\,P_0^2 M_2 \hat{K}_{-1} = -4\,e\,P_0^2 M_2 + \hat{B}. \tag{6.29}$$

Now two of Einstein's field equations [(D.27) and (D.30) both O_3] can be written as equations for $\hat{A}$ and $\hat{B}$ neatly in the complex form

$$\frac{\partial}{\partial \bar{\zeta}}(\hat{A} + i\hat{B}) = -8\,e^2 P_0^2 (L_2 + iM_2)^2 - 2\,e^2(\alpha_2 + i\beta_2) + O_3. \tag{6.30}$$

Here $L_2, M_2, \alpha_2, \beta_2$ are the background functions given by (5.73), (5.88), and (5.89). We note that $\alpha_2 + i\beta_2$ can be expressed as a derivative with respect to $\bar{\zeta}$ given by (5.84). Similarly one can show that

$$P_0^2(L_2 + iM_2)^2 = -2\,\frac{\partial}{\partial\bar{\zeta}}\left(P_0^4(L_2 + iM_2)\,\frac{\partial}{\partial\zeta}(L_2 + iM_2)\right). \tag{6.31}$$

Hence (6.30) can immediately be integrated giving a solution which is free of directional singularities,

$$\begin{aligned}\hat{A} + i\hat{B} = {}& 2\,e^2 P_0^4\frac{\partial}{\partial\zeta}\left\{P_0^{-2}(\alpha_2 + i\beta_2)\right\} + 16\,e^2 P_0^4(L_2 + iM_2)\,\frac{\partial}{\partial\zeta}(L_2 + iM_2)\\ &+ e^2 U(u) + ie^2 V(u) + O_3,\end{aligned} \tag{6.32}$$

where $U(u) = O_0$ and $V(u) = O_0$ are functions of integration. Combining (6.26) with (6.28) and (6.27) with (6.29) and using the solution (6.32) means that we have now determined the functions l_2 and m_2 to be

$$\begin{aligned}l_2 = {}& e\,F_{\rho\sigma}\,k^\rho\,\frac{\partial k^\sigma}{\partial x}\,F_{\mu\nu}\,k^\mu\,v^\nu + \frac{1}{2}e\,F^\rho{}_\mu\,F_{\rho\nu}\,k^\mu\,\frac{\partial k^\nu}{\partial x} - \frac{1}{6}e\,C_{\mu\nu\rho\sigma}\,k^\mu\,v^\nu\,k^\rho\,\frac{\partial k^\sigma}{\partial x}\\ &+ \frac{1}{2}e\,S(u)\,F_{\mu\nu}\,k^\mu\,\frac{\partial k^\nu}{\partial y} + \frac{1}{2}e\,K(u)\,F_{\mu\nu}\,k^\mu\,\frac{\partial k^\nu}{\partial x} - \frac{1}{4}e\,P_0^{-2}U(u) + O_2,\end{aligned} \tag{6.33}$$

$$\begin{aligned}m_2 = {}& e\,F_{\rho\sigma}\,k^\rho\,\frac{\partial k^\sigma}{\partial y}\,F_{\mu\nu}\,k^\mu\,v^\nu + \frac{1}{2}e\,F^\rho{}_\mu\,F_{\rho\nu}\,k^\mu\,\frac{\partial k^\nu}{\partial y} - \frac{1}{6}e\,C_{\mu\nu\rho\sigma}\,k^\mu\,v^\nu\,k^\rho\,\frac{\partial k^\sigma}{\partial y}\\ &- \frac{1}{2}e\,S(u)\,F_{\mu\nu}\,k^\mu\,\frac{\partial k^\nu}{\partial x} + \frac{1}{2}e\,K(u)\,F_{\mu\nu}\,k^\mu\,\frac{\partial k^\nu}{\partial y} - \frac{1}{4}e\,P_0^{-2}V(u) + O_2.\end{aligned} \tag{6.34}$$

One of Maxwell's equations [(D.21) required to be O_2] now gives us directly

$$\hat{K}_1 = F_{\mu\nu}\,k^\mu\,v^\nu + P_0^2\left(\frac{\partial l_2}{\partial x} + \frac{\partial m_2}{\partial y}\right) - 14\,e\,q_2 + O_2, \tag{6.35}$$

and we note that q_2 is given by (5.79).

We now turn our attention to the functions describing the perturbations of the metric tensor. We require $\hat{a}_0, \hat{b}_0, \hat{f}_{-1}$, and $\hat{c}_0$. Defining the variables

$$\hat{\mathcal{A}} = \hat{a}_0 + 4\,P_0^2 L_2\,\hat{K}_0, \qquad \text{and} \qquad \hat{\mathcal{B}} = \hat{b}_0 + 4\,P_0^2 M_2\,\hat{K}_0, \tag{6.36}$$

two of Einstein's equations [(D.28) and (D.31) taken to be O_2] can be written as the one complex equation

$$\frac{\partial}{\partial\bar{\zeta}}(\hat{\mathcal{A}} + i\hat{\mathcal{B}}) = 4\,m\,P_0^2(L_2 + iM_2)^2 + 4\,m\,(\alpha_2 + i\beta_2) + O_2. \tag{6.37}$$

This is integrated in similar fashion to (6.30). The solution should have the property that $P_0^{-1}\hat{a}_0$ and $P_0^{-1}\hat{b}_0$ are free of directional singularities since it is in this way that $\hat{a}_0$ and $\hat{b}_0$ appear in the metric tensor. The solutions of (6.37) therefore that we require are

$$\hat{a}_0 = 2\,e\,P_0^2 a_\rho\, k^\rho\, F_{\mu\nu}\, k^\mu\, \frac{\partial k^\nu}{\partial x} - 2\,m\,P_0^2 F^\rho{}_\mu\, F_{\rho\nu}\, k^\mu\, \frac{\partial k^\nu}{\partial x}$$
$$-4\,m\,P_0^2 F_{\rho\sigma}\, k^\rho\, v^\sigma\, F_{\mu\nu}\, k^\mu\, \frac{\partial k^\nu}{\partial x} - \frac{4}{3} m\, P_0^2 C_{\mu\nu\rho\sigma}\, k^\mu\, v^\nu\, k^\rho\, \frac{\partial k^\sigma}{\partial x}$$
$$+X(u) + O_2, \tag{6.38}$$
$$\hat{b}_0 = 2\,e\,P_0^2 a_\rho\, k^\rho\, F_{\mu\nu}\, k^\mu\, \frac{\partial k^\nu}{\partial y} - 2\,m\,P_0^2 F^\rho{}_\mu\, F_{\rho\nu}\, k^\mu\, \frac{\partial k^\nu}{\partial y}$$
$$-4\,m\,P_0^2 F_{\rho\sigma}\, k^\rho\, v^\sigma\, F_{\mu\nu}\, k^\mu\, \frac{\partial k^\nu}{\partial y} - \frac{4}{3} m\, P_0^2 C_{\mu\nu\rho\sigma}\, k^\mu\, v^\nu\, k^\rho\, \frac{\partial k^\sigma}{\partial y}$$
$$+Y(u) + O_2, \tag{6.39}$$

where $X(u) = O_1$ and $Y(u) = O_1$ are functions of integration. We note that $\hat{a}_0$ and $\hat{b}_0$ possess directional singularities but $P_0^{-1}\hat{a}_0$ and $P_0^{-1}\hat{b}_0$ do not, as required. Another of Einstein's field equations [(D.51) required to be O_3] provides us with the equation

$$\Delta \hat{f}_{-1} - 2\,e^2 a_\mu\, k^\mu + m\, P_0^2 \left(\frac{\partial}{\partial x}\,(P_0^{-2}\hat{a}_{-1}) + \frac{\partial}{\partial y}\left(P_0^{-2}\hat{b}_{-1}\right)\right) = O_3, \tag{6.40}$$

which, using (6.26) and (6.27), can be rewritten as

$$\Delta(\hat{f}_{-1} - e^2 a_\mu\, k^\mu - 2\,m\,e\,F_{\mu\nu}\, k^\mu\, v^\nu) = O_3, \tag{6.41}$$

from which we conclude that

$$\hat{f}_{-1} = e^2 a_\mu\, k^\mu + 2\,m\,e\,F_{\mu\nu}\, k^\mu\, v^\nu + G(u) + O_3, \tag{6.42}$$

with $G(u) = O_2$ a function of integration. Now we obtain directly from an Einstein equation [(D.25) taken to be O_3] the function $\hat{c}_0$. After some considerable simplification it can be written

$$\hat{c}_0 = 1 + 8\,e\,F_{\mu\nu}\, k^\mu\, v^\nu + \Delta Q_2 + 2\,Q_2 - 4\,e^2 C_{\mu\nu\rho\sigma}\, k^\mu\, v^\nu\, k^\rho\, v^\sigma$$
$$+8\,e^2 F^{\mu\nu}\, F_{\mu\nu} + 24\,e^2 (F_{\mu\nu}\, k^\mu\, v^\nu)^2 - 24\,e^2 F^\rho{}_\mu\, F_{\rho\nu}\, k^\mu\, v^\nu$$
$$+\frac{65}{3}\, e^2 F^\rho{}_\mu\, F_{\rho\nu}\, k^\mu\, k^\nu + 16\,e^2 K(u)\, F_{\mu\nu}\, k^\mu\, v^\nu + 4\,e^2 U(u)\, \frac{\partial}{\partial x}(\log P_0)$$
$$+4\,e^2 V(u)\, \frac{\partial}{\partial y}(\log P_0) + O_3. \tag{6.43}$$

The perturbation in the function $\hat{c}_1 = c_1 + O_1$ is now obtained from (D.26) or (D.49) required to be O_2. We shall not need this function in the sequel. The remaining quantities in (D.13)–(D.21) and (D.23)–(D.51) can now be evaluated and their orders of magnitude are indicated in Appendix D. The differential equation for Q_2 emerges from the remaining quantity (D.52) required to be O_3. This expression is lengthy and so we split it into a sum of three terms which we call T_1, T_2, T_3. The first of these terms involves only $\hat{a}_{-1}$ and $\hat{b}_{-1}$ and is given by

$$
\begin{aligned}
T_1 = &-\left(\frac{\partial \hat{a}_{-1}}{\partial x}\right)^2 - \left(\frac{\partial \hat{b}_{-1}}{\partial y}\right)^2 - \frac{1}{2}\left(\frac{\partial \hat{a}_{-1}}{\partial y}\right)^2 - \frac{1}{2}\left(\frac{\partial \hat{b}_{-1}}{\partial x}\right)^2 \\
&- \frac{\partial \hat{a}_{-1}}{\partial y}\,\frac{\partial \hat{b}_{-1}}{\partial x} + 4\,\hat{b}_{-1}\,\frac{\partial \hat{b}_{-1}}{\partial y}\,P_0^{-1}\frac{\partial P_0}{\partial y} + 4\,\hat{a}_{-1}\,\frac{\partial \hat{a}_{-1}}{\partial x}\,P_0^{-1}\frac{\partial P_0}{\partial x} \\
&+2\,\hat{a}_{-1}\frac{\partial \hat{b}_{-1}}{\partial y}\,P_0^{-1}\frac{\partial P_0}{\partial x} + 2\,\hat{b}_{-1}\frac{\partial \hat{a}_{-1}}{\partial x}\,P_0^{-1}\frac{\partial P_0}{\partial y} + 2\,\hat{a}_{-1}\,\frac{\partial \hat{b}_{-1}}{\partial x}\,P_0^{-1}\frac{\partial P_0}{\partial y} \\
&+2\,\hat{b}_{-1}\,\frac{\partial \hat{a}_{-1}}{\partial y}\,P_0^{-1}\frac{\partial P_0}{\partial x} - 2\,(\hat{a}_{-1})^2 P_0\,\frac{\partial^2}{\partial x^2}\,(P_0^{-1}) - 2\,(\hat{b}_{-1})^2 P_0\,\frac{\partial^2}{\partial y^2}\,(P_0^{-1}) \\
&-8\,\hat{a}_{-1}\,\hat{b}_{-1}\,P_0^{-2}\frac{\partial P_0}{\partial x}\,\frac{\partial P_0}{\partial y} - \hat{a}_{-1}\left\{\frac{\partial^2 \hat{a}_{-1}}{\partial x^2} + \frac{\partial^2 \hat{b}_{-1}}{\partial x \partial y}\right\} \\
&-\hat{b}_{-1}\left\{\frac{\partial^2 \hat{b}_{-1}}{\partial y^2} + \frac{\partial^2 \hat{a}_{-1}}{\partial x \partial y}\right\}.
\end{aligned}
\tag{6.44}
$$

To evaluate this we only require the leading terms in $\hat{a}_{-1}$ and $\hat{b}_{-1}$ given in (6.26) and (6.27). By (5.72) these can be written

$$
\hat{a}_{-1} = 2\,e\,P_0 F_{13} + O_2\ , \qquad \text{and} \qquad \hat{b}_{-1} = 2\,e\,P_0 F_{23} + O_2. \tag{6.45}
$$

Direct calculation reveals

$$
\frac{\partial \hat{a}_{-1}}{\partial y} = -\frac{\partial \hat{b}_{-1}}{\partial x} + O_2 \qquad = 2e\left\{\frac{\partial P_0}{\partial y}\,F_{13} - \frac{\partial P_0}{\partial x}\,F_{23} + F_{12}\right\} + O_2, \tag{6.46}
$$

$$
\frac{\partial \hat{a}_{-1}}{\partial x} = \frac{\partial \hat{b}_{-1}}{\partial y} + O_2 \qquad = 2e\left\{\frac{\partial P_0}{\partial x}\,F_{13} + \frac{\partial P_0}{\partial y}\,F_{23} - F_{34}\right\} + O_2, \tag{6.47}
$$

$$
\begin{aligned}
\frac{\partial^2 \hat{a}_{-1}}{\partial x\,\partial y} = &\ e\left\{-P_0^{-1} + 2\,\frac{\partial^2 P_0}{\partial y^2} - 2\,P_0^{-1}\left(\frac{\partial P_0}{\partial x}\right)^2\right\} F_{23} - 2\,e\,P_0^{-1}F_{24} \\
&+2\,e\,P_0^{-1}\frac{\partial P_0}{\partial x}\,\frac{\partial P_0}{\partial y}\,F_{13} + 2\,e\,P_0^{-1}\frac{\partial P_0}{\partial x}\,F_{12} - 2\,e\,P_0^{-1}\frac{\partial P_0}{\partial y}\,F_{34},
\end{aligned}
\tag{6.48}
$$

$$
\begin{aligned}
\frac{\partial^2 \hat{b}_{-1}}{\partial x\,\partial y} = &\ e\left\{-P_0^{-1} + 2\,\frac{\partial^2 P_0}{\partial x^2} - 2\,P_0^{-1}\left(\frac{\partial P_0}{\partial y}\right)^2\right\} F_{13} - 2\,e\,P_0^{-1}F_{14} \\
&+2\,e\,P_0^{-1}\frac{\partial P_0}{\partial x}\,\frac{\partial P_0}{\partial y}\,F_{23} - 2\,e\,P_0^{-1}\frac{\partial P_0}{\partial y}\,F_{12} - 2\,e\,P_0^{-1}\frac{\partial P_0}{\partial x}\,F_{34},
\end{aligned}
\tag{6.49}
$$

where F_{ab} are the components of the external electromagnetic field evaluated on $r = 0$ and referred to the tetrad given by (5.22)–(5.25). Substitution in (6.44) leads to the simplification

$$
T_1 = 4\,e^2\left\{(F_{13})^2 + (F_{23})^2\right\} + 8\,e^2\left\{F_{13}\,F_{14} + F_{23}\,F_{24}\right\} - 8\,e^2(F_{34})^2 + O_3. \tag{6.50}
$$

This is further reduced by noting that

$$(F_{13})^2 + (F_{23})^2 = -F^{\rho}{}_{\mu}\, F_{\rho\nu}\, k^{\mu}\, k^{\nu}, \tag{6.51}$$

$$F_{13}\, F_{14} + F_{23}\, F_{24} = -F^{\rho}{}_{\mu}\, F_{\rho\nu}\, k^{\mu}\, v^{\nu} + \frac{1}{2}\, F^{\rho}{}_{\mu}\, F_{\rho\nu}\, k^{\mu}\, k^{\nu} - (F_{\mu\nu} k^{\mu}\, v^{\nu})^2, \tag{6.52}$$

and

$$F_{34} = F_{\mu\nu} k^{\mu}\, v^{\nu}. \tag{6.53}$$

Hence we can write

$$T_1 = -8\, e^2 F^{\rho}{}_{\mu}\, F_{\rho\nu} k^{\mu}\, v^{\nu} - 16\, e^2 (F_{\mu\nu} k^{\mu}\, v^{\nu})^2. \tag{6.54}$$

The second term T_2 is chosen because it involves the function $\hat{c}_0$ explicitly. It is defined by

$$T_2 = \frac{1}{2}\, \hat{\Delta}\hat{c}_0 - \frac{1}{2}\hat{c}_0\, \hat{P}_0^2 \left(\frac{\partial}{\partial x} \left(\hat{P}_0^{-2} \hat{a}_{-1} \right) + \frac{\partial}{\partial y} \left(\hat{P}_0^{-2} \hat{b}_{-1} \right) \right) - \frac{1}{2} \hat{a}_{-1}\, \frac{\partial \hat{c}_0}{\partial x} - \frac{1}{2} \hat{b}_{-1}\, \frac{\partial \hat{c}_0}{\partial y}. \tag{6.55}$$

Explicit evaluation yields

$$\begin{aligned} T_2 = {} & -6\, e\, F_{\mu\nu}\, k^{\mu}\, v^{\nu} + \frac{1}{2} \Delta(\Delta Q_2 + 2\, Q_2) - 3\, e^2 U(u)\, \frac{\partial}{\partial x} (\log P_0) \\ & -3\, e^2 V(u)\, \frac{\partial}{\partial y} (\log P_0) - 12\, e^2 K(u)\, F_{\mu\nu}\, k^{\mu}\, v^{\nu} + 11\, e^2 C_{\mu\nu\rho\sigma}\, k^{\mu}\, v^{\nu}\, k^{\rho}\, v^{\sigma} \\ & -\frac{59}{3}\, e^2 F^{\mu\nu}\, F_{\mu\nu} - 42\, e^2 (F_{\mu\nu}\, k^{\mu}\, v^{\nu})^2 - 48\, e^2 F^{\rho}{}_{\mu}\, F_{\rho\nu}\, v^{\mu}\, v^{\nu} \\ & -59\, e^2 F^{\rho}{}_{\mu}\, F_{\rho\nu}\, k^{\mu}\, k^{\nu} + \frac{338}{3}\, e^2 F^{\rho}{}_{\mu}\, F_{\rho\nu}\, k^{\mu}\, v^{\nu} + O_3. \end{aligned} \tag{6.56}$$

The final term T_3 consists of the terms remaining in (D.52). It is given by

$$\begin{aligned} T_3 = {} & 3\, m\, P_0^2 \left\{ \frac{\partial}{\partial x} \left(P_0^{-2} \hat{a}_0 \right) + \frac{\partial}{\partial y} (P_0^{-2} \hat{b}_0) \right\} - 4\, \frac{\partial \hat{f}_{-1}}{\partial u} + 12\, \hat{f}_{-1}\, h_0 + 6\, m\, h_0 \\ & -4 e^2 P_0^2 L_2\, \frac{\partial \hat{K}_1}{\partial x} - 4 e^2 P_0^2 M_2\, \frac{\partial \hat{K}_1}{\partial y} - 2\, P_0^2 \left\{ \left(\frac{\partial \hat{K}_0}{\partial x} \right)^2 + \left(\frac{\partial \hat{K}_0}{\partial y} \right)^2 \right\} \\ & +8\, m\, P_0^2\, L_2\, \frac{\partial \hat{K}_0}{\partial x} + 8\, m\, P_0^2\, M_2\, \frac{\partial \hat{K}_0}{\partial y} - 4\, P_0^2 L_2 \frac{\partial \hat{K}_{-1}}{\partial x} - 4\, P_0^2 M_2 \frac{\partial \hat{K}_{-1}}{\partial y} \\ & -4\, P_0^2 \frac{\partial K_1}{\partial x}\, \frac{\partial \hat{K}_{-1}}{\partial x} - 4\, P_0^2 \frac{\partial K_1}{\partial y}\, \frac{\partial \hat{K}_{-1}}{\partial y} \\ & -\frac{5}{2} e^2\, P_0^2 \left\{ \frac{\partial}{\partial x} \left(P_0^{-2} a_1 \right) + \frac{\partial}{\partial y} \left(P_0^{-2} b_1 \right) \right\}. \end{aligned} \tag{6.57}$$

When this is evaluated with the functions derived above it results in (using $h_0 = a_\mu\,k^\mu$ again for convenience)

$$\begin{aligned}
T_3 = {}& 6\,m\,h_0 + 2\,e^2 a_\mu\,a^\mu + 14\,e^2 h_0^2 - 4\,e^2 \dot{h}_0 - 4\,\dot{G} + 12\,h_0\,G - 8\,m\,e\,\dot{F}_{\mu\nu}\,k^\mu\,v^\nu \\
& + \left(6\,m^2 + \frac{5}{3}\,e^2\right) F^{\mu\nu}\,F_{\mu\nu} - (12\,m^2 + 5\,e^2)\,C_{\mu\nu\rho\sigma}\,k^\mu\,v^\nu\,k^\rho\,v^\sigma \\
& -6\,m\,X(u)\,\frac{\partial}{\partial x}(\log P_0) - 6\,m\,Y(u)\,\frac{\partial}{\partial y}(\log P_0) \\
& +(50\,e^2 - 36\,m^2)\,(F_{\mu\nu}\,k^\mu\,v^\nu)^2 + 12\,e^2 F^\rho{}_\mu\,F_{\rho\nu}\,v^\mu\,v^\nu \\
& - \left(36\,m^2 + \frac{14}{3}\,e^2\right) F^\rho{}_\mu\,F_{\rho\nu}\,k^\mu\,v^\nu + (18\,m^2 + 5\,e^2)\,F^\rho{}_\mu\,F_{\rho\nu}\,k^\mu\,k^\nu \\
& +O_3.
\end{aligned} \tag{6.58}$$

6.3 The equations of motion

Now the approximate field equation giving us a differential equation for Q_2 is obtained from $T_1 + T_2 + T_3 = O_3$ with T_1, T_2, T_3 given by (6.54), (6.56), and (6.58). Making use of the unit space-like vector field p^μ introduced prior to (6.16) we can write the result in the form

$$-\frac{1}{2}\Delta(\Delta Q_2 + 2\,Q_2) = A_0 + A_1 + A_2 + O_3, \tag{6.59}$$

where

$$A_0 = -4\,\dot{G} - \frac{16}{3}e^2 F^\rho{}_\mu\,F_{\rho\nu}\,v^\mu\,v^\nu, \tag{6.60}$$

which is an $l = 0$ spherical harmonic,

$$\begin{aligned}
A_1 = {}& 6\,m\,a_\mu\,p^\mu - 6\,e\,F_{\mu\nu}\,p^\mu\,v^\nu - 4\,e^2 h^\nu_\mu \dot{a}_\nu\,p^\mu - 8\,m\,e\,\dot{F}_{\mu\nu}\,p^\mu\,v^\nu \\
& -8\,e^2 h^\lambda_\mu\,F^\rho{}_\lambda\,F_{\rho\nu}\,p^\mu\,v^\nu + 12\,G\,a_\mu\,p^\mu - 12\,e^2 K(u)\,F_{\mu\nu}\,p^\mu\,v^\nu \\
& -3\,(e^2 U(u) + 2\,m\,X(u))\,\frac{\partial}{\partial x}(\log P_0) \\
& -3\,(e^2 V(u) + 2\,m\,Y(u))\,\frac{\partial}{\partial y}(\log P_0),
\end{aligned} \tag{6.61}$$

which is an $l = 1$ spherical harmonic, and

$$A_2 = 6\,(e^2 - 2m^2)\,\chi_{(1)} - 4\,(2\,e^2 + 9m^2)\,\chi_{(2)} + 18(m^2 - 3e^2)\,\chi_{(3)} + 18e^2\chi_{(4)}, \tag{6.62}$$

which is an $l = 2$ spherical harmonic. Any $l = 2$ spherical harmonic can be written

$$\chi = D_{\mu\nu}(u)\,k^\mu\,k^\nu, \tag{6.63}$$

with

$$D_{\mu\nu} = D_{\nu\mu}\,, \qquad D_{\mu\nu}\,v^\nu = 0\,, \qquad \eta^{\mu\nu}\,D_{\mu\nu} = 0. \tag{6.64}$$

In (6.62) each $\chi_{(a)}$ for $a = 1, 2, 3, 4$ has the form

$$\chi_{(a)} = {}_{(a)}D_{\mu\nu}\,k^\mu\,k^\nu, \tag{6.65}$$

with

$${}_{(1)}D_{\mu\nu} = C_{\mu\rho\nu\sigma}\,v^\rho\,v^\sigma, \tag{6.66}$$

$${}_{(2)}D_{\mu\nu} = F_{\mu\rho}v^\rho\,F_{\nu\sigma}v^\sigma - \frac{1}{3}h_{\mu\nu}F^\rho{}_\lambda\,F_{\rho\sigma}\,v^\lambda\,v^\sigma, \tag{6.67}$$

$${}_{(3)}D_{\mu\nu} = F^\rho{}_\lambda\,F_{\rho\sigma}\,h^\lambda_\mu\,h^\sigma_\nu - \frac{1}{3}h_{\mu\nu}\,h^{\lambda\sigma}\,F^\rho{}_\lambda\,F_{\rho\sigma}, \tag{6.68}$$

$${}_{(4)}D_{\mu\nu} = a_\mu\,a_\nu - \frac{1}{3}h_{\mu\nu}\,a_\lambda\,a^\lambda. \tag{6.69}$$

The last two terms in (6.61) can be simplified by noting that from (5.15) and (5.26) we can write

$$\frac{\partial}{\partial x}(\log P_0) = c_\mu\,k^\mu = c_\mu\,p^\mu\,, \qquad \frac{\partial}{\partial y}(\log P_0) = d_\mu\,k^\mu = d_\mu\,p^\mu, \tag{6.70}$$

with

$$c_\mu = \left(\frac{1}{2}v^1, \frac{1}{2}(v^3 - v^0), 0, -\frac{1}{2}v^1\right) \quad \text{and} \quad d_\mu = \left(\frac{1}{2}v^2, 0, \frac{1}{2}(v^3 - v^0), -\frac{1}{2}v^2\right), \tag{6.71}$$

covariant vectors (in coordinates (X^μ)) defined along $r = 0$ and everywhere orthogonal to v^μ. Provided $A_0 = O_3$, (6.59) can be integrated without the introduction of directional singularities to read

$$\Delta Q_2 + 2\,Q_2 = A_1 + \frac{1}{3}A_2 + O_3, \tag{6.72}$$

up to the addition of an arbitrary function of u which makes a trivial contribution to Q_2. For Q_2 to be free of directional singularities we must have $A_1 = O_3$ for all p^μ such that $p^\mu\,v_\mu = 0$ and this leads to the equations of motion

$$\begin{aligned} m a_\mu &= e\,F_{\mu\nu}\,v^\nu + \frac{2}{3}e^2 h^\nu_\mu \dot{a}_\nu + \frac{4}{3}e^2 h^\lambda_\mu\,F^\rho{}_\lambda\,F_{\rho\nu}\,v^\nu \\ &\quad + \frac{4}{3}m\,e\,\dot{F}_{\mu\nu}\,v^\nu - 2\,G\,a_\mu + 2\,e^2 K(u)\,F_{\mu\nu}\,v^\nu + \hat{U}(u)\,c_\mu + \hat{V}(u)\,d_\mu + O_3, \end{aligned} \tag{6.73}$$

where we have written $\hat{U} = (e^2 U(u) + 2\,m\,X(u))/2 = O_2$ and $\hat{V} = (e^2 V(u) + 2\,m\,Y(u))/2 = O_2$. Putting $A_1 = O_3$ and discarding the geometrically trivial homogeneous solution of (6.72) we see that $Q_2 = -A_2/12$ and this describes smooth non-trivial perturbations of the wave fronts near the black hole as required. $G(u)$ is given by (6.60) with $A_0 = O_3$ and thus can be written as a definite integral for which we would naturally choose $(-\infty, u]$ as the range of integration resulting in $-2\,G\,a_\mu$ in (6.73) contributing to a "tail term". We can write $\hat{U}\,c_\mu + \hat{V}\,d_\mu = \Omega_{\mu\nu}\,v^\nu$ with $\Omega_{\mu\nu} = -\Omega_{\nu\mu}$ vanishing except for $\Omega_{13} = \Omega_{01} = \hat{U}/2$ and $\Omega_{23} = \Omega_{02} = \hat{V}/2$. Define $\omega_{\mu\nu}(u) = -\omega_{\nu\mu}(u)$ by $\dot{\omega}_{\mu\nu} = 2\,e^2 K(u)\,F_{\mu\nu} + \Omega_{\mu\nu}$. If we now make the 1-parameter family of infinitesimal Lorentz transformations on the unit tangent vector, $v^\mu \to \bar{v}^\mu = v^\mu - (4/3) m\,e\,F^\mu{}_\nu\,v^\nu - (1/m)\omega^\mu{}_\nu\,v^\nu$, we find that, after dropping the bar on v and its derivatives with respect to u, *the equations of motion* take the form

$$m a_\mu = e\,F_{\mu\nu}\,v^\nu + \frac{2}{3} e^2 h^\nu_\mu \dot{a}_\nu + \frac{4}{3} e^2 h^\nu_\mu\,F^\rho{}_\nu\,F_{\rho\lambda}\,v^\lambda + \mathcal{T}_\mu + O_3, \tag{6.74}$$

where

$$\mathcal{T}_\mu = \frac{e}{m}\,\{\omega_\rho{}^\nu\,F_{\nu\mu} - \omega_\mu{}^\nu\,F_{\nu\rho} - 2\,G\,F_{\mu\rho}\}\;v^\rho = O_2. \tag{6.75}$$

Since $\omega_{\mu\nu}$ can be written as an integral, whose range we would naturally take to be $(-\infty, u]$ as in the case of G above, we see that (6.75) is a "tail term" in a set of equations of motion formally of the DeWitt and Brehme (1960) form. Direct mathematical comparison with the equations of motion in DeWitt and Brehme (1960) is highly non-trivial since the equations derived there have involved the removal of an infinite term while deriving (6.74) clearly does not involve such a procedure. Hence we say that (6.74) fits the pattern of the equations obtained in DeWitt and Brehme (1960).

6.4 Observations and prospects

Perhaps the most significant aspect of the approach presented here is the fact that the equations of motion (6.74) of a small charged black hole moving in an external electromagnetic and gravitational field have been derived from the vacuum Einstein–Maxwell field equations without encountering infinities. It is also important to note that the external fields have been introduced in Chapter 5 as a solution of the Einstein–Maxwell vacuum field equations.

The second term on the right-hand side of (6.74) is the electromagnetic radiation reaction 4-force. The third term on the right-hand side of (6.74) can be written in terms of the electromagnetic energy tensor $E_{\mu\nu}(u)$ of the background space-time, calculated on $r = 0$ in coordinates (X^μ). Using the background field equations on $r = 0$, $R_{\mu\nu} = 2\,E_{\mu\nu}$, this term can be written in terms of the background Ricci tensor components $R_{\mu\nu}$ as $(2/3)e^2 h^\rho_\mu\,R_{\rho\nu}\,v^\nu$. This is the 4-force due to the external field and it differs from that derived by Hobbs (1968) by a factor of 2. The external electromagnetic field in Hobbs (1968) is *not* required to satisfy the Einstein–Maxwell vacuum field equations and so the background Ricci tensor appearing there does not have the external electromagnetic field as source. Thus in effect half of the contribution to the external 4-force is neglected.

The "tail term" in the form (6.75) vanishes if the external electromagnetic field vanishes. It involves three arbitrary functions (of u) $K, \hat{U}, \hat{V}$ with the latter two obtained from U, V, X, Y as indicated following (6.73). The mathematical origin of these functions as functions of integration arising in the determination of $\hat{K}_{-1}, l_2$ and m_2, required for the perturbed 4-potential, and of $\hat{a}_0$ and $\hat{b}_0$ required for the perturbed metric tensor, is clear from this work. In addition the tail term indirectly involves the two vector fields c^μ and d^μ orthogonal to v^μ. It is a topic for further study to understand physically the role of these arbitrary functions and of these unique space-like vectors. Had the tail term not been so explicit we could not identify these arbitrary functions and space-like vectors for further consideration. By comparison the tail term obtained in DeWitt and Brehme (1960) is an integral whose integrand is expressed in terms of functions appearing in the Hadamard form of the Green function of the vector wave equation. Very little is known explicitly about the form of such functions. The DeWitt–Brehme tail term has proved quite intractable to analyze due to the fact that the Green function is not known in closed form. This makes a comparison with us extremely difficult.

The original calculations of Dirac (1938) were carried out in Minkowskian space-time. Our results do not specialize to those of Dirac since our background space-time is an Einstein–Maxwell space-time and if it is flat then the external electromagnetic field vanishes. In this case (6.74) gives $ma_\mu = O_2$ and when this is substituted into the surviving radiation reaction 4-force the latter is absorbed into the O_3-error. Hence we obtain geodesic motion, and in particular no "runaway" motion, at this level of approximation in this theory. This is the Reissner–Nordström special case of a small charged black hole in the absence of external fields.

When the small black hole is specialized to a small Schwarzschild black hole moving in an external vacuum gravitational field the equations of motion simplify to

$$ma_\mu = O_3, \tag{6.76}$$

which corresponds to geodesic motion in this approximation. In this case the external Weyl tensor becomes the external Riemann tensor and although it has no influence at this level of approximation on the equations of motion it does enter the perturbed metric tensor, for example, via the functions $\hat{a}_0$ and $\hat{b}_0$ given by (6.38) and (6.39). If the equations of motion are calculated in this case with an O_4-error we still expect to obtain geodesic motion. However if the O_4-term is explicitly calculated we expect it to be quadratic in the Riemann tensor calculated on the world-line $r = 0$ in the background space-time, and thereby geodesic motion is violated at the level of approximation in which the error is O_5. Confirming these speculations is a considerable future challenge.

7 Gravitational physics of few-body systems

A subject closely related to equations of motion is the gravitational physics of few-body systems. First we review choreographic solutions for the N-body problem in Newton's theory of gravity. A solution is called choreographic if each massive particle moves periodically in a single closed orbit. One such choreographic solution is a stable *figure-eight* orbit for a three-body system. We describe connections between the choreographic solutions and the gravitational waves that are predicted by Einstein's theory of general relativity and thus general relativistic choreography is discussed. General relativistic effects cause, for example to binary orbits, a complicated flower-like pattern by the periastron advance. It is not trivial whether general relativistic effects admit a choreographic solution such as the figure-eight. At least at the first and second post-Newtonian orders (neglecting dissipative effects starting at the second and half post-Newtonian order) the tricky figure-eight remarkably remains true.

7.1 Choreographic solutions for the N-body problem

The three-body problem in Newtonian gravity is one of the longstanding problems in astronomy and classical physics [see for example Danby (1988)]. In 1765 Leonhard Euler discovered a collinear solution, and Joseph-Louis Lagrange found an equilateral triangle solution in 1772. It is impossible to describe all the solutions to the three-body problem even for the $1/r$ potential. In his celebrated paper, Poincarè proved that we cannot analytically obtain all the solutions, and the number of new solutions is increasing (Marchal, 1990). Therefore the three-body problem remains unsettled even for Newtonian gravity. The Newtonian N-body problem admits *choreographic* solutions, which have attracted increasing interest since a choreographic figure-eight solution to the three-body problem was found, first by Moore (1993), and re-discovered with its existence proof by Chenciner and Montgomery (2000). A solution is called choreographic in celestial mechanics if every massive particle moves periodically in a single closed orbit. An equal-mass binary in circular motion is a trivial example of a choreographic solution.

The simplest periodic solutions for the three-body problem were discovered by Euler (1765) and Lagrange (1772). Euler's solution is a collinear solution in which the masses are collinear at every instant with the same ratios of their distances. Lagrange's solution is an equilateral triangle solution in which each mass moves in an ellipse in such a way that the triangle formed by the three bodies revolves. Built out of Keplerian ellipses, they are the only explicit solutions, for which each mass moves on an ellipse. A choreographic solution for which three bodies move periodically in a single figure-eight orbit was found first by topological arguments and numerical computations (Moore, 1993). This solution is stable in Newtonian theory (Simo, 2002; Galan *et al.*, 2002). The figure-eight seems unique up to scaling and rotation according to all numerical investigations, and indeed its uniqueness has been recently proven (Montgomery, 2005). It is a one-parameter family which is determined uniquely by the total energy. Many efforts have been made to reveal properties such as convexity in the figure-eight solution (Kapela and Zgliczynski, 2003; Fujiwara and Montgomery, 2005). Furthermore it has been demonstrated numerically that fourth, sixth, or eighth-order polynomials cannot result in the figure-eight solution (Simo, 2002). Nevertheless no analytic expressions in closed form for the figure-eight shape have been found up to now. It is an open issue to find its analytic form.

For simplicity we assume a 3-body system with each mass equal to m. Without loss of generality the orbital plane is taken as the $x - y$ plane. The position of each mass is denoted by (x_A, y_A) for $A = 1, 2, 3$. Figure 7.1 shows the figure-eight orbit, where we take the initial condition as $(x_1, y_1) = (-x_2, -y_2) = (0.970, -0.243)$ and $(\dot{x}_3, \dot{y}_3) = (-2\dot{x}_1, -2\dot{y}_1) = (-2\dot{x}_2, -2\dot{y}_2) = (-0.932, -0.865)$, where a dot denotes the time derivative (Simo, 2002). When one mass arrives at the knot (center) of the figure-eight, ℓ is defined as one half of the separation between the remaining two masses. It is convenient to use ℓ instead of a distance between the knot and the apoapsis because the inertial moment is expressed simply as $2m\ell^2$. Clearly this system has no Killing

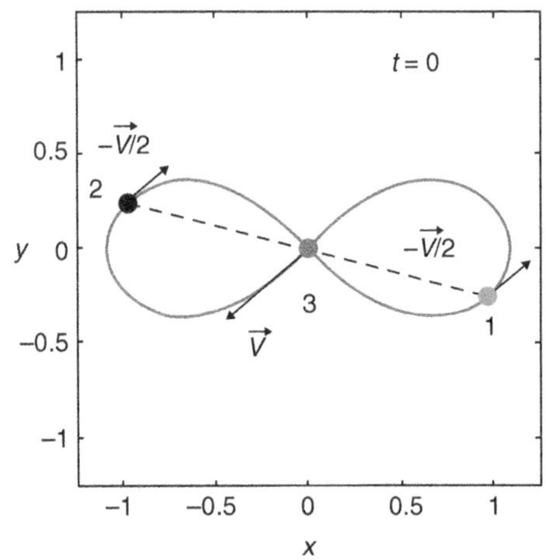

Fig. 7.1 Three masses in a figure-eight at the initial time. Each mass is labelled by 1, 2 and 3. The initial velocity of each mass is indicated with arrows. The distance between 1 and 3 is denoted by ℓ. In this plot we take $\ell = 1$ (Chiba *et al.*, 2007).

vector as can be seen in Fig. 7.1. The orbital period is estimated as (Chenciner and Montgomery, 2000; Chiba *et al.*, 2007)

$$T = 6.32591398\sqrt{\frac{\ell^3}{Gm}},$$
$$\approx 10^4 \left(\frac{M_\odot}{m}\right)^{1/2} \left(\frac{\ell}{R_\odot}\right)^{3/2} \text{s}, \tag{7.1}$$

where G denotes the Newtonian gravitational constant and $M_\odot$ and $R_\odot$ are the solar mass and radius respectively. Figure 7.2 shows a configuration of the three masses after one half of the period. The mass labelled 3 starts at the knot and sweeps out of the left-hand side of the figure-eight during the first half of the period.

7.2 Gravitational waves from few-body systems

Gravitational wave physics represents one of the great challenges of fundamental physics. Gravitational waves will open a new window in astronomy [see for example Misner *et al.* (1973), Centrella (2003), and Mio (2006*a*, 2006*b*)]. The tiny ripples in a curved space-time generated by accelerated masses have not been detected directly. A (quasi-)periodic source is the most promising candidate for the first detection. Hence (quasi-)periodic sources have been studied for the last few decades in numerous articles involving a single star in rotation/oscillation (Kokkotas and Schmidt, 1999; Stergioulas, 2003) or double stars in binary motion (Asada and Futamase, 1997*b*; Blanchet, 2006). In particular event rates of an inspiraling and finally merging compact binary (Curran and Lorimer, 1995; VandenHeuvel and Lorimer, 1996; Arzoumanian *et al.*, 1999; Kalogera *et al.*, 2001; Kim *et al.*, 2003; Burgay *et al.*, 2003) are discussed

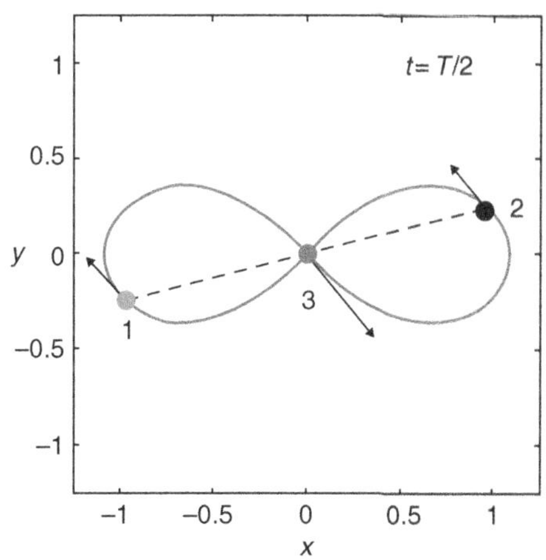

Fig. 7.2 A figure-eight at $t = T/2$. The velocity of each mass is indicated with arrows (Chiba *et al.*, 2007).

by taking account of large-scale detectors such as LIGO, VIRGO, GEO600, TAMA300, and LISA. Furthermore gravitational waves generated by a binary system have been intensively studied by perturbations such as the post-Newtonian approximation (Asada and Futamase, 1997*b*; Blanchet, 2006; Faye *et al.*, 2004; Berti *et al.*, 2006), or by full numerical simulations for neutron stars (Shibata, 1999) and for black holes (Pretorius, 2005; Campanelli *et al.*, 2006*b*; Baker *et al.*, 2006; Diener *et al.*, 2006).

On the other hand less attention has been paid to N-body gravitating systems because of their chaotic orbits when $N \geq 3$. There are existing works which are limited in the sense that they study a binary system affected by the existence of a third body, namely perturbations induced by the third object (Ioka *et al.*, 1998; Wardell, 2002; Campanelli *et al.*, 2006*a*) or test particles in orbits around a black hole (Detweiler, 1978; Nakamura and Oohara, 1983; Nakamura and Sasaki, 1981; Sasaki and Tagoshi, 2003; Teukolsky, 1973). In this section we discuss whether or not N-body systems can generate (quasi-)periodic gravitational waves.

In a wave zone the gravitational waves denoted by h_{ij}^{TT} become asymptotically

$$h_{ij}^{TT} = \frac{2G\ddot{Q}_{ij}}{rc^4} + O\left(\frac{1}{r^2}\right), \tag{7.2}$$

where r is a distance from a source (Misner *et al.*, 1973; Blanchet, 2006). Here TT means the transverse $\left(h_{ij}^{TT} n^i = 0\right)$ and traceless $\left(h_{ii}^{TT} = 0\right)$ part of h_{ij} and n^i denotes the unit vector in the direction of propagation of the gravitational waves. The reduced quadrupole Q_{ij} is defined as $Q_{ij} = I_{ij} - \delta_{ij} I_{kk}/3$. The quadrupole moment I_{ij} for N masses is expressed as $I_{ij} = \sum_{A=1}^{N} m_A x_A^i x_A^j$, where m_A denotes the A-th mass at the location x_A^i, and i and j run from 1 to 3 ($x^1 = x, x^2 = y, x^3 = z$). Figure 7.3 shows a volcano-shaped waveform from the figure-eight solution.

The loss rate of the energy (E) and the angular momentum (L^i) is given by the quadrupole formula as Misner *et al.* (1973) and Blanchet (2006),

$$\frac{dE}{dt} = -\frac{G}{5c^5}\left\langle I_{ij}^{(3)} I_{ij}^{(3)} \right\rangle, \tag{7.3}$$

$$\frac{dL^i}{dt} = -\frac{2G}{5c^5}\epsilon^{ijk}\left\langle I_{ja}^{(2)} I_{ak}^{(3)} \right\rangle. \tag{7.4}$$

Here the brackets denote the time averaging and ϵ^{ijk} denotes the three-dimensional Levi-Civita symbol defined as the completely antisymmetric unit tensor by $\epsilon^{123} = 1$. Because $z_A^3 = 0$ for each mass, dL^x/dt and dL^y/dt vanish. In the case of the figure-eight solution we can show numerically that dL^z/dt vanishes. Why do the emitted waves carry away no total angular momentum? Each body is likely to lose its orbital angular momentum by gravitational waves. This apparent paradox happens because the total angular momentum of this 3-body system vanishes, while the angular momentum of each body does not necessarily vanish, as seen from Fig. 7.1. As a result gravitational waves carry away the angular momentum of each body, while it is impossible for this system to lose its total angular momentum. For instance in Fig. 7.1, at the initial time, the particle labelled 3 has no orbital angular momentum. The particle 1 has the orbital angular momentum, which is carried away by gravitational waves. The same is true of particle 2 but with the opposite sign. Therefore these changes in

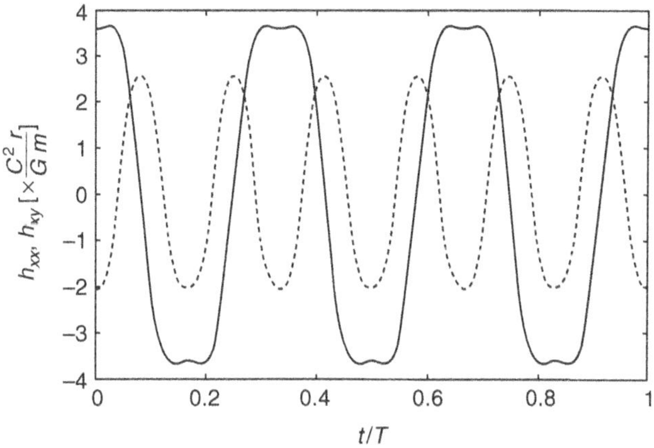

Fig. 7.3 Gravitational waves from three masses in a figure-eight orbit. The dashed curve and the solid one denote h_{xx}^{TT} and h_{xy}^{TT} respectively. This plot is in contrast to that of an equal-mass binary, for which gravitational waves are expressed as a simple sine curve with one maximum and one minimum. On the other hand there is a small valley between the nearest maximum and a small ridge between the nearest minimum in Figure 7.3. At the bottom of the valley and the top of the ridge one of the three masses passes the knot of the figure-eight, corresponding to Figs. 7.1 and 7.2 respectively (Chiba *et al.*, 2007).

the orbital angular momentum cancel each other. This cancellation is expressed as $\sum_{A=1}^{3} \dot{L}_A = 0$, where L_A denotes the angular momentum of each body. The absence of angular momentum radiation as a consequence of such a cancellation is in contrast to the case of the head-on collision of two non-spinning black holes (Smarr, 1979; Price and Pullin, 1994). In the head-on collision case each black hole moves without any orbital angular momentum with respect to the common center of mass. In fact the angular momentum of each black hole is not carried away ($\dot{L}_B = 0$ for each black hole). Clearly no total angular momentum is radiated. The energy loss rate is estimated as (Chiba *et al.*, 2007)

$$\frac{dE}{dt} = 1.2 \times 10^{19} \left(\frac{m}{M_\odot}\right)^5 \left(\frac{R_\odot}{\ell}\right)^5 \text{ erg/s}. \tag{7.5}$$

As a result the radiation reaction time scale is evaluated as (Chiba, Imai, and Asada, 2007)

$$\begin{aligned} t_{GW} &\equiv \frac{E}{dE/dt}, \\ &= 0.13 \left(\frac{M_\odot}{m}\right)^3 \left(\frac{\ell}{R_\odot}\right)^4 \text{ Gyr}, \end{aligned} \tag{7.6}$$

which gives a rough estimate of the collision time. It is worth mentioning that the dependence of T, dE/dt, and t_{GW} on m and ℓ is the same as that on the mass and separation of a binary, as implied also by a dimensional analysis.

Let us consider a "compact" system of three black holes just before merging. To do so we take ℓ as a much shorter distance, say a dozen Schwarzschild radii for each mass. Then the orbital period of this system is of the order of a few milli-seconds. The frequency of the associated gravitational waves becomes a few kHz, around which the large-scale interferometric detectors (LIGO, VIRGO, GEO600, TAMA300) are most sensitive. We obtain the amplitude of the gravitational waves to be (Chiba *et al.*, 2007)

$$h_{ij}^{TT} \sim 10^{-17} \left(\frac{m}{10M_\odot}\right) \left(\frac{10M_{sch}}{\ell}\right) \left(\frac{10\text{kpc}}{r}\right), \tag{7.7}$$

where M_{sch} denotes the Schwarzschild radius.

As matter of fact we have to take account of relativistic corrections such as the back-reaction on the evolution of the orbit. If the system is secularly stable against the gravitational radiation, one might see probably a shrinking ($\dot{\ell} < 0$) figure-eight orbit as a consequence of decreasing the total energy ($\dot{E} < 0$)and keeping the vanishing total angular momentum ($L = 0$). Indeed a chirp signal caused by the back-reaction at the inspiraling phase plays a crucial role in the gravitational wave detection. Furthermore numerical relativity seems necessary for evolving the system in its merging phase after the last stable orbit.

Finally we comment on the possibility of gravitational waves from 3-body systems in a 'periodic' orbit such as a figure-eight. Obviously the possibility seems extremely low though it is quite uncertain. A quick order-of-magnitude estimate is made in Chiba *et al.* (2007). As a new outcome of binary–binary scattering the figure-eight orbit was discussed by Heggie (2000) for presenting a way of detecting such an orbit in numerical computations. According to the numerical result the probability of the formation of figure-eight orbits is a tiny fraction of one percent. Here we assume, rather optimistically, one per galaxy which implies 10^6 figure-eight systems within 100 Mpc. If ℓ distributes randomly with its mean of O(1 AU), we use Eq. (7.6) to obtain $t_{GW} \sim 10^8$ Gyr for $\ell \sim 1$ AU. As a result merging event rates are of the order of 10^{-2} event/Gyr within 100 Mpc, which is extremely low. Therefore the above estimate suggests that figure-eight sources are not likely to be observed.

7.3 General relativistic figure-eight solution

The theory of general relativity is currently the most successful gravitational theory describing the nature of space and time, and is well confirmed by observations. In particular it has passed "classical" tests, such as the deflection of light, the perihelion shift of Mercury and the Shapiro time delay, and also a systematic test using the remarkable binary pulsar "PSR 1913+16" (Will, 1994). It is worthwhile examining the three-body (or more generally, N-body) problem in general relativity. N-body dynamics in general relativistic gravity plays important roles in astrophysics. For instance the formation of massive black holes in star clusters is tackled mostly by Newtonian N-body simulations [see for example Zwart *et al.* (2004)]. However it is more difficult to work this out in general relativity compared with Newtonian gravity because Einstein's equations are so much more complicated (Misner *et al.*, 1973) [even for a two-body system (Pati and Will, 2000; Faye *et al.*, 2004; Blanchet, 2006; Asada and Futamase, 1997*b*)].

Future space astrometric missions such as SIM, GAIA, and JASMINE require a general relativistic modelling of the solar system within an accuracy of a micro arc-second (Klioner, 2003). In addition gravitational wave astronomy needs an accurate modelling of compact binaries. A binary plus a third body has been studied for perturbations of gravitational waves induced by the third body (Ioka *et al.*, 1998; Wardell, 2002; Campanelli *et al.*, 2006*a*; Gultekin *et al.*, 2006). We describe below, in the framework of general relativity, a recent development in choreographic solutions to the three-body problem.

For simplicity we assume a three-body system with each mass equal to m. Without loss of generality the orbital plane is taken again as the x–y plane. Here we take ℓ as 100, while it was unity in the previous section. Why $\ell = 100$? This is because we will treat the post-Newtonian correction in terms of the ratio between the mass and the separation such as ℓ (Imai *et al.*, 2007). In our case the ratio m/ℓ is 0.01 (i.e. the post-Newtonian correction is about one percent). In the equations of motion the second post-Newtonian (2PN) corrections of the order of $(m/\ell)^3$ can be safely neglected if m/ℓ is very small, say 10^{-8}. In this case however the Newtonian and relativistic orbits will be indistinguishable. In order to exaggerate the difference between the two orbits we choose m/ℓ as 0.01 for which the first post-Newtonian (1PN) terms are several dozen times larger than 2PN ones. In our computations, which are not long-time integrations over, say, thousands of orbital periods, we can assume that 1PN terms are enough to bring major relativistic effects to bear on the dynamics.

The position of each mass (m_A) is denoted by (x_A, y_A) for $A = 1, 2, 3$. Figure 7.4 shows the figure-eight orbit, where we take the initial condition as $\boldsymbol{\ell} \equiv (x_1, y_1) = (-x_2, -y_2) = (97.00, -24.31)$, $(x_3, y_3) = (0, 0)$ and $\boldsymbol{V}_{\text{Newton}} \equiv (\dot{x}_3, \dot{y}_3) = (-2\dot{x}_1, -2\dot{y}_1) = (-2\dot{x}_2, -2\dot{y}_2) = (-0.09324, -0.08647)$, where a dot denotes the time derivative (Simo, 2002). When one mass arrives at the knot (center) of the figure-eight, $\ell \equiv |\boldsymbol{\ell}|$ is one half of the separation between the remaining two masses. It is convenient to use ℓ instead of a distance between the knot and the apoapsis because the inertial moment is expressed simply as $2m\ell^2$. The orbital period is given by Eq. (7.1)

In Chapters 2 and 3 we saw how the motion of massive bodies is predominantly described by the Newtonian equations of motion. In order to include the dominant general relativistic effects, we must take account of the terms at the first post-Newtonian order. Thus the motion of the massive bodies obeys the Einstein *et al.* (1938) (EIH) equations of motion. The EIH equations are derived from the first post-Newtonian Lagrangian as (Landau and Lifshitz, 1975)

$$
\begin{aligned}
\mathcal{L} = & \frac{1}{2}\sum_A m_A v_A^2 + \frac{1}{2}\sum_A \sum_{B \neq A} \frac{m_A m_B}{r_{AB}} + \frac{1}{8}\sum_A m_A v_A^4 \\
& -\frac{1}{4}\sum_A \sum_{B \neq A} \frac{m_A m_B}{r_{AB}} \left[7(\boldsymbol{v}_A \cdot \boldsymbol{v}_B) - 6v_A^2 + (\boldsymbol{v}_A \cdot \boldsymbol{n}_{AB})(\boldsymbol{v}_B \cdot \boldsymbol{n}_{AB})\right] \\
& -\frac{1}{2}\sum_A \sum_{B \neq A} \sum_{C \neq A} \frac{m_A m_B m_C}{r_{AB} r_{AC}},
\end{aligned}
\tag{7.8}
$$

where we define

$$\boldsymbol{r}_{AB} \equiv \boldsymbol{r}_A - \boldsymbol{r}_B, \tag{7.9}$$

$$r_{AB} \equiv |\boldsymbol{r}_{AB}|, \tag{7.10}$$

$$\boldsymbol{n}_{AB} \equiv \frac{\boldsymbol{r}_{AB}}{r_{AB}}. \tag{7.11}$$

Figure 7.4 shows an orbit of a body starting at the Newtonian initial condition described above (Imai *et al.*, 2007). In Fig. 7.4 a figure-eight orbit does not seem to survive at the 1PN order. However this is not the case. We should note that the initial condition at the 1PN order does not necessarily coincide with that for Newtonian gravity. We will thus carefully examine the initial condition by taking account of 1PN corrections. For this purpose we assume that both the linear and angular momenta are zero (i.e. $\boldsymbol{P} = 0$ and $\boldsymbol{L} = 0$).

We should remember $\boldsymbol{v}_1 = \boldsymbol{v}_2 = -(\boldsymbol{v}_3)/2$ for the Newtonian figure-eight, for which both the total linear and angular momenta are zero. The changes in $\boldsymbol{v}_1$ and $\boldsymbol{v}_2$ are expressed by using two vectors $\boldsymbol{v}_3$ and $\boldsymbol{\ell}$ which are linearly independent. Hence the initial velocity of each mass is parameterized as in Imai *et al.* (2007):

$$\boldsymbol{v}_1 = k\boldsymbol{V} + \xi\frac{m}{\ell^3}(\boldsymbol{V}\cdot\boldsymbol{\ell})\boldsymbol{\ell}, \tag{7.12}$$

$$\boldsymbol{v}_2 = k\boldsymbol{V} + \xi\frac{m}{\ell^3}(\boldsymbol{V}\cdot\boldsymbol{\ell})\boldsymbol{\ell}, \tag{7.13}$$

$$\boldsymbol{v}_3 = \boldsymbol{V}, \tag{7.14}$$

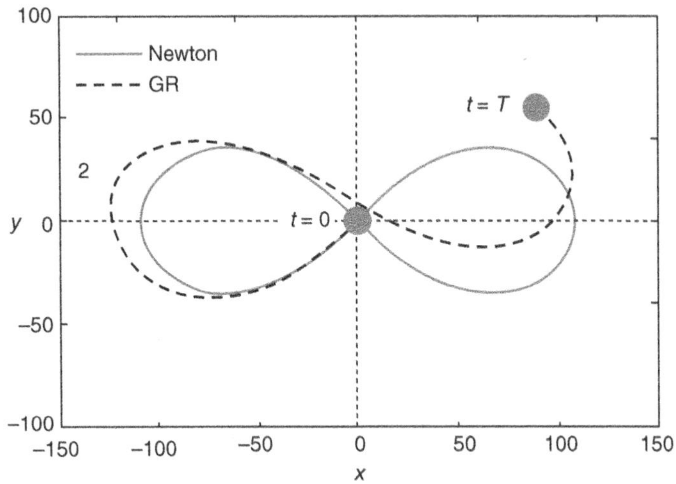

Fig. 7.4 Figure-eights starting at the Newtonian initial condition. The solid curve denotes a figure-eight orbit in Newtonian gravity. The dashed curve denotes a trajectory of one mass following the EIH equations of motion under the same Newtonian initial condition (Chiba *et al.*, 2007).

where k is expressed as

$$k = -\frac{1}{2} + \alpha |\boldsymbol{V}|^2 + \beta \frac{m}{\ell}. \tag{7.15}$$

Here the 1PN terms involve either $|\boldsymbol{V}|^2$ or m/ℓ. If $\boldsymbol{P} = 0$ and $\boldsymbol{L} = 0$ then there is no need for $|\boldsymbol{V}|^2$ in front of $\boldsymbol{\ell}$ in (7.12) and (7.13) as shown below. The linear and angular momenta are calculated from the first post-Newtonian Lagrangian (Landau and Lifshitz, 1975). The total angular momentum identically vanishes owing to the parameterizations given by (7.12)–(7.15). The total linear momentum becomes

$$\boldsymbol{P} = m \left[\left(2\alpha + \frac{3}{8} \right) |\boldsymbol{V}|^2 + \left(2\beta - \frac{1}{4} \right) \frac{m}{\ell} \right] \boldsymbol{V} + 2m \left(\xi - \frac{1}{8} \right) \frac{m^2}{\ell^3} (\boldsymbol{V} \cdot \boldsymbol{\ell}) \boldsymbol{\ell}. \tag{7.16}$$

Here we impose the condition $\boldsymbol{P} = 0$ at 1PN order. Then we can determine the 1PN coefficients as

$$\alpha = -\frac{3}{16}, \tag{7.17}$$

$$\beta = \frac{1}{8}, \tag{7.18}$$

$$\xi = \frac{1}{8}. \tag{7.19}$$

Up to this point $\boldsymbol{V}$ is arbitrary. Next we determine $\boldsymbol{V}$. The initial velocity of the particles can be different from that for Newtonian gravity. The post-Newtonian terms affect both the magnitude and direction of the velocity. Therefore by using two linearly independent vectors $\boldsymbol{\ell}$ and $\boldsymbol{V}_{\text{Newton}}$ we parameterize the initial velocity as

$$\boldsymbol{V} = \left(1 + \delta \frac{m}{\ell} \right) \boldsymbol{V}_{\text{Newton}} + \eta \frac{m}{\ell} \frac{\boldsymbol{\ell}}{\ell} \left(\boldsymbol{V}_{\text{Newton}} \cdot \frac{\boldsymbol{\ell}}{\ell} \right), \tag{7.20}$$

where it is sufficient to express 1PN corrections in terms of either m/ℓ or $|\boldsymbol{V}_{\text{Newton}}|^2$ in numerical computations, though both are necessary for analytic calculations of $\boldsymbol{P} = 0$ and $\boldsymbol{L} = 0$. For convenience we choose m/ℓ in (7.20). By numerically performing trial and error iterations until achieving a periodic orbit we find that

$$\delta = -3.3, \tag{7.21}$$

$$\eta = -3.7. \tag{7.22}$$

Here the iterative computations are done until we find the values of δ and η for which the three masses simultaneously return to their initial positions. Our procedure is as follows: if and only if one particle returns to the neighborhood of its initial position (the origin for the particle labelled 3) within the positional deviation of 0.01, we measure how far the remaining two particles are from their initial positions at the same moment when the particle is closest to the initial position. For (7.21) and (7.22) the sum of the squared distances is minimized as approximately 0.1. This is sufficient for $\ell = 100$ and $m/\ell = 0.01$ because expected positional shifts after one cycle are of the order of unity or more. For instance such a shift exceeds ten for $\alpha = \beta = \xi = \delta = \eta = 0$ in

Fig. 7.4. We integrate the motion over ten cycles to confirm the periodicity. After ten periods the solution found comes back to the same point within a deviation of ± 1. The numerical computation gives the orbital period as

$$T_{\rm GR} \approx \left(1 + \frac{6m}{\ell}\right) \times T_{\rm Newton}. \tag{7.23}$$

The relativistic figure-eight appears to be stable, in the sense that we recognize eight-like orbits that resemble figures of eight for slightly different values of δ and η. It is a future project to analyze the long-time stability of the relativistic figure-eight.

Figure 7.5 shows that a figure-eight orbit is still closed even after including the dominant general relativistic effects. In Fig. 7.5 we can recognize an asymmetric difference between the Newtonian figure-eight orbit and the general relativistic (GR) one. The deviation is partly fiducial because the principal axes of the GR figure-eight orbit are not along the x and y axes. That is $\boldsymbol{\ell}$, which defines the direction of the initial position of a particle with respect to the principal axes, changes slightly at the 1PN order. The axes are inclined at 0.012 radians with respect to those of the Newtonian figure-eight orbit. Now we choose the x-axis as the principal axis for both the Newtonian figure-eight orbit and the general relativistic one. After choosing the principal axes Fig. 7.6 shows a general relativistic choreographic solution at the first post-Newtonian order. The solution recovers line symmetry with respect to the x and y axes. There are no significant differences in the velocity between the Newtonian and GR figure-eight orbits. One may notice that the 2PN terms have been neglected.

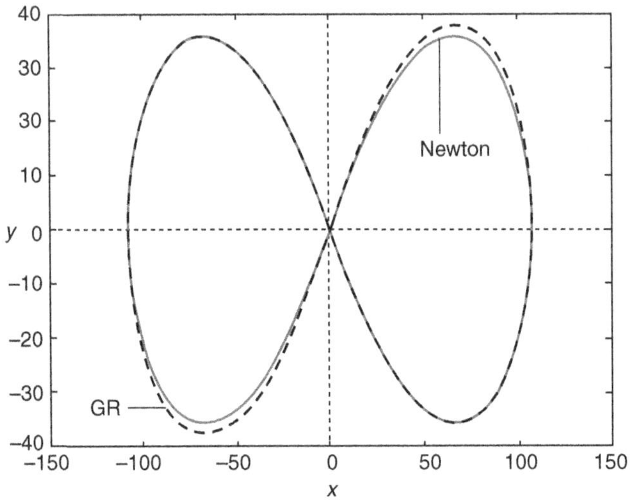

Fig. 7.5 Figure-eight orbits. The solid curve denotes a figure-eight orbit in Newtonian gravity. The dashed curve denotes a figure-eight orbit at the 1PN order of general relativity (Chiba *et al.*, 2007).

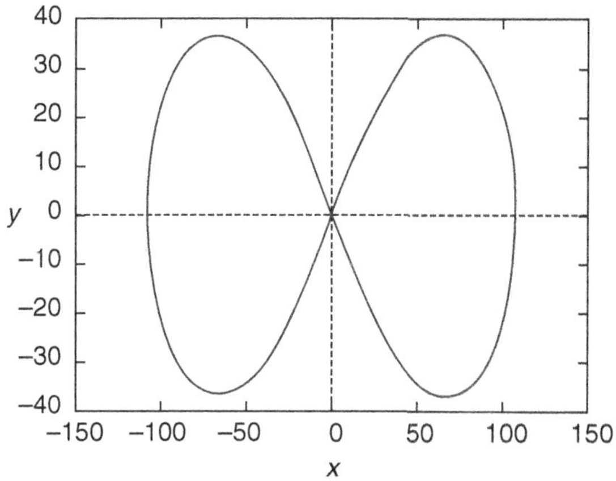

Fig. 7.6 Relativistic figure-eight. The principal axes of the orbit are chosen as x and y axes (Chiba *et al.*, 2007).

It would be safer to choose m/ℓ for instance as 10^{-8} and then on the figures to exaggerate the differences between the Newtonian and post-Newtonian solutions in order to make them visible.

Appendix A
Far zone contribution

A formal retardation expansion gives a divergent integral when we evaluate $h^{\mu\nu}$. Schematically the field variable $h(\tau, x^i)$ at the field point (τ, x^i) is given by (omitting indices for simplicity)

$$h(\tau, x^i) \sim \int d^3y \, \frac{f(\tau - \epsilon|\vec{x} - \vec{y}|, \vec{y})}{|\vec{x} - \vec{y}|} \sim \sum_n^\infty \frac{(-1)^n}{n!} \epsilon^n \int d^3y \, |\vec{x} - \vec{y}|^{n-1} \frac{d^n}{dt^n} f(t, \vec{y}). \quad \text{(A.1)}$$

This integral is divergent if we set the upper bound of the integral to infinity. In our formalism we take the upper bound of the integral as $\sim \mathcal{R}/\epsilon$ and keep the $\mathcal{R}$ dependence in the field. In the last step of the derivation we let $\mathcal{R}$ go to infinity if and only if this procedure gives a finite result at least up to the relevant post-Newtonian order. At a higher order of the retardation expansion approximation however, some terms must have a $\ln \mathcal{R}$ and/or a $\mathcal{R}^n (n > 0)$ dependence. In this case we cannot let $\mathcal{R}$ become infinite. To solve this problem it is important to realize that the field variable at (τ, x^i) consists of two contributions: the retarded integral in the near zone (the near zone contribution) and the retarded integral outside the near zone (the far zone contribution). Since $\mathcal{R}$ is introduced artificially we expect that the far zone contribution to the near zone field must have some $\mathcal{R}$ dependent terms which cancel out completely the $\mathcal{R}$ dependent terms in the near zone. This expectation is correct and was proved to any post-Newtonian order by Pati and Will (2000) using their formalism. They have shown that the total field (which is the sum of the near zone contribution and the far zone contribution) is finite and independent of $\mathcal{R}$. We give here the calculation of the far zone contribution to the 3PN equations of motion following Pati and Will (2000), Will (1999), Will and Wiseman (1996) [see also Pati and Will (2002), Will (2005)] and establish their result that the far zone field does not have any influence on the equations of motion up to and including the 3PN order.

For the near zone field we write $h^{\mu\nu}$ as

$$h^{\mu\nu}_{N(C)} = h^{\mu\nu}_{N(N)} + h^{\mu\nu}_{N(F)} + h^{\mu\nu}_{\mathrm{H}}, \quad \text{(A.2)}$$

$$h^{\mu\nu}_{N(N)} = 4 \int_{N=\{y:|y| \le \mathcal{R}/\epsilon\}} d^3y \, \frac{\Lambda^{\mu\nu}(\tau - \epsilon|\vec{x} - \vec{y}|, y^k; \epsilon)}{|\vec{x} - \vec{y}|}, \quad \text{(A.3)}$$

$$h^{\mu\nu}_{N(F)} = 4 \int_{F=\{y:|y| > \mathcal{R}/\epsilon\}} d^3y \, \frac{\Lambda^{\mu\nu}(\tau - \epsilon|\vec{x} - \vec{y}|, y^k; \epsilon)}{|\vec{x} - \vec{y}|}, \quad \text{(A.4)}$$

where $h_N^{\mu\nu}$ is the field in the near zone, $h_{N(N)}^{\mu\nu}$ is the near zone integral contribution to the near zone field, and $h_{N(F)}^{\mu\nu}$ is the far zone integral contribution to the near zone field. The term $h_{\rm H}^{\mu\nu}$ is given by (2.67). Our task here is to evaluate $h_{N(F)}^{\mu\nu}$ to the 3PN order. This means that we should derive $h_F^{\mu\nu}$ to lower order because the integrand $\Lambda^{\mu\nu}$ in the retarded integral for $h_{N(F)}^{\mu\nu}$ consists of $h_F^{\mu\nu}$. To simplify the presentation here we shall omit the bookkeeping parameter ϵ although it should be remembered that $\vec{v}$ is of order ϵ and m of order ϵ^2. We evaluate the near zone contribution to the far zone field using a multipole expansion,

$$h_{F(N)}^{\mu\nu} = 4\int_N \frac{d^3y}{|\vec{x}-\vec{y}|}\Lambda^{\mu\nu}(t-|\vec{x}-\vec{y}|,\vec{y}) = 4\sum_{l=0}\frac{(-1)^l}{l!}\partial_{K_l}\left(\frac{1}{r}M^{K_l\mu\nu}(u)\right), \tag{A.5}$$

where K_l is a collective multi-index, $r=|\vec{x}|$ is the distance from the near zone center to the field point and $u=t-r$. The near zone multipole moments $M^{K_l\mu\nu}$ are defined by

$$M^{K_l\mu\nu}(u) \equiv \int_N d^3y\,\Lambda^{\mu\nu}(u,\vec{y})\,y^{K_l}. \tag{A.6}$$

Next the far zone contribution to the far zone field point can be evaluated as follows. The key idea is to introduce u instead of the time coordinate t with the result that the radial integration is transformed into a temporal integration from past infinity to $u=t-r$ thus

$$\begin{aligned} h_{F(F)}^{\mu\nu}(t,\vec{x}) &= 4\int_F d^3x'\,\frac{\Lambda^{\mu\nu}(t-|\vec{x}-\vec{x}'|,\vec{x}')}{|\vec{x}-\vec{x}'|}, \\ &= 4\int_{-\infty}^{u} du'\oint_F \frac{\Lambda^{\mu\nu}(u'+r',\vec{x}')}{t-u'-\vec{n}'\cdot\vec{x}}[r'(u',\Omega')]^2\,d\Omega, \end{aligned} \tag{A.7}$$

where

$$r'(u',\Omega') = \frac{(t-u')^2-r^2}{2(t-u'-\vec{n}'\cdot\vec{x})}. \tag{A.8}$$

We then make the STF decomposition of the integrand $\Lambda^{\mu\nu}$ in the retarded integral as $\Lambda^{\mu\nu}\sim f_{B,L}r^{-B}n^{\langle L\rangle}$, where $\vec{n}=\vec{r}/r$. The integrand in (A.7) becomes a sum of terms, each of which consists of some algebraic combinations of near zone multipole moments, defined by (A.6), multiplied by terms which explicitly depend on t, u', and Ω'. Roughly speaking the strategy is to integrate by parts many times, each time increasing the number of the time derivatives of the multipole moments, and assume that the system is sufficiently stationary in the past so that the contributions from past infinity disappear. We then have for the far zone contribution to the far zone field point

$$h^{\mu\nu}_{F(F)}(u,x^i) = \sum_{\substack{B\neq 2\\ l}} \left(\frac{2}{r}\right)^{B-2} n^{\langle L\rangle} \sum_{q=0} \mathcal{D}^q_{B,L}(z)\, r^q\, \frac{d^q f_{B,L}(u)}{du^q}$$
$$+\frac{n^{\langle L\rangle}}{r} \int_0^\infty f_{2,L}(u-s) Q_L\left(1+\frac{s}{r}\right)$$
$$+n^{\langle L\rangle} \sum_{q=0} \mathcal{D}^q_{2,L}(z)\, r^q\, \frac{d^q f_{2,L}(u)}{du^q}, \tag{A.9}$$

with $z = \mathcal{R}/r$,

$$\mathcal{D}^q_{B,L}(z) = \frac{(-1)^q}{q!} \int_1^{1+2z} \frac{(\zeta-1)^q}{(\zeta^2-1)^{B-2}} A_{B,L}(\zeta,\alpha)\, d\zeta - \sum_{p=0}^{q} k^{(q-p+1)}_{B,L}(1+2z) \frac{(-2z)^p}{p!}, \tag{A.10}$$

for $B \neq 2$ and

$$\mathcal{D}^q_{2,L}(z) = \frac{(-1)^{q+1}}{2q!} \int_1^{1+2z} (\zeta-1)^q\, d\zeta \int_{-1}^{1-\alpha} \frac{\mathrm{P}_l(y)}{\zeta - y}\, dy, \tag{A.11}$$

for $B = 2$. Other quantities in the above equations are given by

$$A_{B,L}(\zeta,\alpha) \equiv \frac{1}{2} \int_{1-\alpha}^{1} \frac{\mathrm{P}_l(y)}{\zeta - y}\, dy, \tag{A.12}$$
$$\alpha \equiv (\zeta-1)(\zeta+1-2z)/(2z), \tag{A.13}$$
$$\frac{dk^{(m)}_{B,L}(\zeta)}{d\zeta} = k^{(m-1)}_{B,L}(\zeta) \qquad \text{for } m \geq 1, \tag{A.14}$$
$$k^{(0)}_{B,L}(\zeta) \equiv \frac{A_{B,L}(\zeta,2)}{(\zeta^2-1)^{B-2}} \qquad \text{for } m \geq 1. \tag{A.15}$$

Here P_l and Q_l are the Legendre functions of the first and second kind. The variable α represents an angular defect due to the fact that the far zone integral does not cover the whole space-time on account of the near zone. The function $k(t)$ and the retarded time multi-derivatives of the STF coefficient $f_{B,L}$ come from the recursive integrations by parts. We note that in terms of the expansion parameter ϵ the variable $z = \mathcal{R}/r$ satisfies $z < \epsilon$ for the field point in the far zone. Combining the near zone and the far zone contributions to the far zone field point we have, to the required order,

$$h_F^{00} = 4\left[\frac{P_N^0}{r} - \partial_k\left(\frac{D_N^k(u)}{r}\right) + \frac{1}{2}\partial_{kl}\left(\frac{I_N^{kl}(u)}{r}\right) - \frac{1}{6}\partial_{klm}\left(\frac{I_N^{klm}(u)}{r}\right)\right]$$
$$+7\left(\frac{P_N^0}{r}\right)^2, \tag{A.16}$$

$$h_F^{0i} = 4\left[\frac{P_N^i(u)}{r} - \partial_k\left(\frac{J_N^{ki}(u)}{r}\right) + \frac{1}{2}\partial_{kl}\left(\frac{J_N^{kli}(u)}{r}\right)\right], \tag{A.17}$$

$$h_F^{ij} = 4\left[\frac{Z_N^{ij}(u)}{r} - \partial_k\left(\frac{Z_N^{kij}(u)}{r}\right) + \frac{1}{2}\partial_{kl}\left(\frac{Z_N^{klij}(u)}{r}\right)\right]$$
$$+\left(\frac{P_N^0}{r}\right)^2 n^{ij}, \tag{A.18}$$

where $P_N^0 = M^{00}$, $P_N^i = M^{0i}$, $D_N^i = M^{i00}$, $I_N^{ij} = M^{ij00}$, $J_N^{ki} = M^{k0i}$, $J_N^{kli} = M^{kl0i}$, $Z_N^{ij} = M^{ij}$, $Z_N^{kij} = M^{kij}$, and $Z_N^{klij} = M^{klij}$.

Next we evaluate the far zone contribution to the near zone field point. A transformation of the retarded integral (A.7) is again used and similar arguments to those following (A.7) lead to the formulae,

$$h^{\mu\nu}_{N(F)}(t, x^i) = \sum_{\substack{B\neq 2\\ l}}\left(\frac{2}{r}\right)^{B-2} n^{\langle L\rangle}\sum_{q=0}\mathcal{E}_{B,L}(z)^q\, r^q\,\frac{d^q f_{B,L}(t)}{dt^q}$$
$$+\frac{n^{\langle L\rangle}}{r}\int_0^\infty f_{2,L}(u-s)\,\mathrm{Q}_l\left(1+\frac{s}{r}\right)$$
$$+n^{\langle L\rangle}\sum_{q=0}\mathcal{E}^q_{2,L}(z)\,r^q\,\frac{d^q f_{2,L}(t)}{dt^q}, \tag{A.19}$$

with

$$\mathcal{E}^q_{B,L}(z) = \frac{(-1)^q}{q!}\int_{2z-1}^{2z+1}\frac{\zeta^q}{(\zeta^2-1)^{B-2}}A_{B,L}(\zeta,\alpha)\,d\zeta - \sum_{p=0}^{q}k_{B,L}^{(q-p+1)}(1+2z)\frac{(-1-2z)^p}{p!}, \tag{A.20}$$

for $B \neq 2$ and

$$\mathcal{E}^q_{2,L}(z) = \frac{(-1)^{q+1}}{q!}\left\{\int_1^{1+2z}\zeta^q\mathrm{Q}_l(\zeta)\,d\zeta + \frac{1}{2}\int_{2z-1}^{2z+1}\zeta^q\,d\zeta\int_{-1}^{1-\alpha}\frac{\mathrm{P}_l(y)}{\zeta-y}\,dy\right\}, \tag{A.21}$$

for $B = 2$. Evaluating the coefficients $\mathcal{E}^q_{B,L}$ we finally have

$$
\begin{aligned}
h^{\tau\tau}_{N(F)}(\tau, x^i) &= \epsilon^{10} \left[-8P^{\tau}_{N} n^{\langle kl \rangle} \int_1^\infty {}^{(4)}I^{kl}_N(t - r\zeta)\, Q_2(\zeta)\, d\zeta \right. \\
&\quad - \frac{8}{3} P^{\tau}_{N} \int_1^\infty {}^{(4)}I^{kk}_N(t - r\zeta)\, Q_0(\zeta)\, d\zeta \\
&\quad \left. + \frac{4}{3} P^{\tau}_{N} {}^{(4)}I^{kl}_N(t) n^{\langle kl \rangle} + \frac{8}{3} P^{\tau}_{N} {}^{(4)}I^{kk}_N(t) \left(1 + \ln\left(\frac{\mathcal{R}}{\epsilon r}\right)\right)\right] \\
&\quad + O(\epsilon^{11}), \qquad\qquad (A.22) \\
h^{\tau i}_{N(F)} &= O(\epsilon^9), \qquad\qquad (A.23) \\
h^{kk}_{N(F)} &= \epsilon^8 \left[-16 P^{\tau}_{N} \int_1^\infty {}^{(2)}Z^{k}_{N\,k}(t - r\zeta)\, Q_0(\zeta)\, d\zeta \right. \\
&\quad \left. + 16 P^{\tau}_{N} {}^{(2)}Z^{k}_{N\,k} \left(1 + \ln\left(\frac{\mathcal{R}}{\epsilon r}\right)\right)\right] \\
&\quad + O(\epsilon^9), \qquad\qquad (A.24)
\end{aligned}
$$

where we have reintroduced the bookkeeping parameter ϵ and Z^{ij}_N is written in terms of the near zone quadrupole moment as $Z^{ij}_N = (1/2)d^2 I^{ij}_N/d\tau^2$. We note that the fields $h^{\tau\tau}$ of $O(\epsilon^{10})$ and $h^{\mu i}$ of $O(\epsilon^8)$ are the 3PN fields in our formalism. Finally, assuming that in the distant past the binary was sufficiently stationary we find that the far zone contribution becomes a function of time only at the 3PN order. It turns out that only the spatial derivatives of those 3PN fields contribute to the 3PN equations of motion. Thus the far zone contribution does not affect the equations of motion up to 3PN order inclusive.

Appendix B
Effects of extendedness of stars

We derive here the spin–orbit coupling force, the quadrupole–orbit coupling force, the spin–spin coupling force, and the spin geodesic precession equation to the lowest order.

The order-counting of the multipole couplings is done in the following way. The magnitudes of a star's mass multipole moment and current multipole moment of order l are roughly (mass) $\times$ (radius of the star)^l and (mass) $\times$ (radius of the star)^l $\times$ (velocity of steller internal motion) respectively. Since we assume slow stellar rotation, for which the velocity of stellar internal motion is $O(\epsilon)$, and the strong field point particle limit, for which the radius of the star is $O(\epsilon^2)$, the mass multipole moment and the current multipole moment are of order $O(\epsilon^{2l+2})$ and $O(\epsilon^{2l+3})$ respectively. For example the spin–orbit coupling force which takes the form of (mass) $\times$ (orbital velocity) $\times$ (spin) appears at $O(\epsilon^4)$, that is at 2PN order and not at the usual 1.5PN order where rapid stellar rotation is assumed. In summary the structure of the equations of orbital motion can be written schematically as

$$
\begin{aligned}
ma^i = {} & F^i_{\mathrm{Newton}} + \epsilon^2 F^i_{1\mathrm{PN}} + \epsilon^4 F^i_{2\mathrm{PN}} + \epsilon^4 F^i_{\mathrm{SO}} + \epsilon^4 F^i_{\mathrm{QO}} + \epsilon^5 F^i_{\mathrm{RR}} \\
& + \epsilon^6 F^i_{3\mathrm{PN}} + \epsilon^6 F^i_{1\mathrm{PN\,SO}} + \epsilon^6 F^i_{1\mathrm{PN,QO}} + \epsilon^6 F^i_{\mathrm{OO}} + \epsilon^6 F^i_{\mathrm{SS}} + \epsilon^6 F^i_{\mathrm{TO}} \\
& + O(\epsilon^7).
\end{aligned} \tag{B.1}
$$

Here F^i_{Newton}, $F^i_{1\ \mathrm{PN}}$, $F^i_{2\ \mathrm{PN}}$, F^i_{RR}, and $F^i_{3\ \mathrm{PN}}$ are the Newtonian force, the 1PN force, the 2PN force, the 2.5PN radiation reaction force, and the 3PN force respectively. F^i_{SO} and $F^i_{1\ \mathrm{PNSO}}$ are the spin–orbit coupling force and its 1PN correction, while F^i_{QO} and $F^i_{1\ \mathrm{PN\ QO}}$ are the quadrupole–orbit coupling force and its 1PN correction. F^i_{OO}, F^i_{SS}, and F^i_{TO} are the octupole–orbit coupling force, spin–spin coupling force, and tidal–orbit coupling force respectively. We call the coupling between temporal derivatives of the quadrupole moment and the orbital motion the *tidal–orbit coupling*.

Though formally the effects of the 1PN spin–orbit coupling, the 1PN quadrupole–orbit coupling, the octupole–orbit coupling and the tidal–orbit coupling appear up to 3PN order in our ordering, we focus our attention on the lowest-order spin–orbit coupling, the spin–spin coupling and the quadrupole–orbit coupling forces. The 1PN spin–orbit coupling force was derived by Tagoshi, Ohashi, and Owen (2001).

Before investigating the multipole–orbit coupling forces it is worth noticing that our definition of multipole moments is operational and that the relation between these multipole moments and the *intrinsic* multipole moments of a star has not been given. We address briefly this problem. For an appropriate frame of reference for the definition of multipole moments it is natural to use a frame attached to the star which is nonrotating with respect to an asymptotic inertial frame [see Brumberg and Kopeikin (1989) for the case of an earth–satellite system in the solar system]. If we do not define the multipole moments in an appropriate frame then, for example, an apparent quadrupole moment would be produced by Lorentz contraction caused by the orbital motion of the star and an apparent spin would be produced by the Thomas precession. One realization of an appropriate frame is to use the (generalized) Fermi normal coordinates (Ashby and Bertotti, 1986). To derive the spin–orbit coupling force in the same form as in previous works of Damour (1982), Kidder (1995), and Thorne and Hartle (1985), it is sufficient to assume the coordinate transformation of $\Lambda_N^{\tau\tau}$ in the near zone coordinates to $\Lambda_A^{\mu'\nu'}$ in the (generalized) Fermi normal coordinates in the following implicit form [see Ashby and Bertotti (1986), Brumberg and Kopeikin (1989), Damour *et al.* (1991, 1992, 1993, 1994) and also Section B.0.4 below],

$$\Lambda_N^{\tau\tau} = \left(\Gamma^{\tau}_{A\tau'}\right)^2 \Lambda_A^{\tau'\tau'} + 2\epsilon^2 \Gamma^{\tau}_{A\tau'} \Gamma^{\tau}_{A\underline{i}'} \Lambda_A^{\tau'\underline{i}'} + O(\epsilon^2), \tag{B.2}$$

with $\Gamma^{\tau}_{A\tau} = 1 + O(\epsilon^2)$ and $\Gamma^{\tau}_{Ai} = v_A^i + O(\epsilon^2)$, reminiscent of a Lorentz transformation. The ϵ^2 in front of the second term arises from the body zone coordinates rescaling $\left(x_A^i = \epsilon^2 \alpha_A^i\right)$. An explicit expression of $\Gamma^{\mu}_{A\nu}$ is not required for our purpose. To 2PN order, which is a sufficient order to derive the lowest-order spin–orbit, spin–spin, and quadrupole–orbit coupling forces, the transformation changes only ${}_8h^{\tau\tau}$,

$$h^{\tau\tau} = 4\epsilon^4 \sum_{A=1,2} \left[\frac{P_A^{\tau}}{r_A} + \epsilon^2 \frac{r_A^i}{r_A^3}\left(d_A^i + \epsilon^2 M_A^{ik} v_A^k\right)\right] + \dots, \tag{B.3}$$

where the dots stand for irrelevant terms and

$$d_A^i = \epsilon^2 \int_{B_A} d^3\alpha_A \left(\Gamma^{\tau}_{A\tau'}\right)^2 \alpha_A^{\underline{i}} \Lambda_A^{\tau'\tau'}, \tag{B.4}$$

$$m_A^{ij} = 2\epsilon^4 \int_{B_A} d^3\alpha_A\, \Gamma^{\tau}_{A\tau'} \alpha_A^{[\underline{i}} \Lambda_A^{\underline{j}']\tau'} + O(\epsilon) = M_A^{ij} + O(\epsilon), \tag{B.5}$$

where M_A^{ij} is given by (2.88). Next we consider the χ part of the multipole moments. We have used $\Lambda_N^{\mu\nu} = \Theta_N^{\mu\nu} + \chi_N^{\mu\nu\alpha\beta}{}_{,\alpha\beta}$ in the definition of the multipole moments since we could not evaluate the χ parts of multipole moments separately except for some low-order moments. However we have to take into account the fact that the χ parts of our higher multipole moments can affect the equations of motion for two point masses. An obvious example can be found from the definition of the energy and the mass. It is *natural* to define the mass as a volume integral of $\Theta_N^{\tau\tau}$. In fact the χ part of the energy $P_{A\chi}^{\tau}$ appears at 2PN order. As another example, the χ part of the dipole moment can be evaluated directly as

$$D_{A\chi}^i = \epsilon^2 \int_{B_A} d^3\alpha_A\, \alpha_A^{\underline{i}} \chi_N^{\tau\tau\alpha\beta}{}_{,\alpha\beta} = \epsilon^4 \frac{175\, m_1^3\, m_2\, r_{12}^i}{18\, r_{12}^3} + O(\epsilon^5). \tag{B.6}$$

Thus if we were to define the center of mass of the star, not by $D_A^i = 0$ but by $D_{A\Theta}^i = 0$, then the form of the equations of motion for the two point masses would change. We note that we can evaluate the χ part of the spin by virtue of its antisymmetric nature and the result is $M^{ij}_{A\chi} = O(\epsilon^3)$. Hence $M^{ij}_{A\chi}$ does not affect the 3PN equations of motion for two point masses up to 3PN order even if we treat $M^{ij}_{A\chi}$ separately from $M^{ij}_{A\Theta}$.

Finally we list the relevant $h^{\mu\nu}$ and Q^i up to the required order to derive the lowest-order spin–orbit, spin–spin, and quadrupole–orbit coupling forces together with the spin geodesic precession equation,

$$\begin{aligned} h^{\tau\tau} &= 4\epsilon^6 \sum_{A=1,2} \frac{r^k_A}{r^3_A} D^k_A + 6\epsilon^8 \sum_{A=1,2} \frac{r^k_A r^l_A}{r^5_A} I^{kl}_A, \\ &= 4\epsilon^6 \sum_{A=1,2} \frac{r^k_A}{r^3_A} d^k_A + 6\epsilon^8 \sum_{A=1,2} \frac{r^k_A r^l_A}{r^5_A} I^{kl}_A + 4\epsilon^8 \sum_{A=1,2} \frac{r^k_A}{r^3_A} M^{ki}_A v^i_A \\ &\quad + O(\epsilon^9), \end{aligned} \tag{B.7}$$

$$ {}_6h^{\tau i} = 2 \sum_{A=1,2} \frac{r^k_A}{r^3_A} M^{ki}_A, \tag{B.8}$$

$$ {}_6h^{ij} = 4 \sum_{A=1,2} \frac{r^k_A}{r^3_A} M^{k(i}_A v^{j)}_A, \tag{B.9}$$

$$ Q^i_1 = \epsilon^4 \frac{2m_2 M^{ik}_1}{3r^3_{12}} r^k_{12}. \tag{B.10}$$

B.0.1 Spin–orbit coupling force

It is well known that the definition of the dipole moment of a star, which we equate to zero to determine the center of mass of the star, affects the appearance of the spin–orbit coupling force [see for example Kidder (1995)]. If one chooses $d^i_A = 0$, as in Itoh *et al.* (2000), then the corresponding spin–orbit coupling force takes the usual form [Damour (1982), Kidder (1995), Thorne and Hartle (1985)],

$$\begin{aligned} F^i_{\rm 1SO}|_{d^i_A=0} &= -\epsilon^4 \frac{V^k}{r^3_{12}} \left[\left(2m_1 M^{il}_2 + m_2 M^{il}_1\right) \Delta^{lk} + 2\left(m_1 M^{lk}_2 + m_2 M^{lk}_1\right)\Delta^{li}\right], \\ &= \epsilon^4 \frac{m_1}{r^3_{12}} \left[6(\vec{s}_2 \times \vec{n}_{12}) \cdot \vec{V} n^i_{12} + 4\vec{s}_2 \times \vec{V} - 6\vec{s}_2 \times \vec{n}_{12}(\vec{n}_{12}\cdot\vec{V})\right] \\ &\quad + \epsilon^4 \frac{m_2}{r^3_{12}} \left[6(\vec{s}_1 \times \vec{n}_{12}) \cdot \vec{V} n^i_{12} + 3\vec{s}_1 \times \vec{V} - 3\vec{s}_1 \times \vec{n}_{12}(\vec{n}_{12}\cdot\vec{V})\right], \end{aligned} \tag{B.11}$$

where $\Delta^{ij} \equiv \delta^{ij} - 3n^i_{12}n^j_{12}$ and $\times$ in this section denotes the vector product of Euclidean three-vectors. The spin vector s^i_A is defined by

$$s_A^i \equiv \frac{1}{2}\epsilon_{ijk}M_A^{jk}, \tag{B.12}$$

where ϵ_{ijk} is the three-dimensional Levi-Civita permutation symbol.

B.0.2 Spin–spin coupling force

It is straightforward to derive the spin–spin coupling force. The definition of the center of mass does not change the form of the spin–spin coupling force, as one would expect. We evaluate the surface integral

$$F^i_{1\mathrm{SS}} = -\oint_{\partial B_1} dS_j \,{}_{10}\left[(-g)t^{ij}_{\mathrm{LL}}\right] + \ldots, \tag{B.13}$$

with ${}_{10}\left[(-g)t^{ij}_{\mathrm{LL}}\right] = \delta^{ij}{}_6h^{\tau k,l}{}_6h^{\tau}{}_{[k,l]} + 2{}_6h^{\tau[k,i]}{}_6h^{\tau j}{}_{,k} + 2{}_6h^{\tau[k,j]}{}_6h^{\tau i}{}_{,k} + \ldots.$ The result is

$$\begin{aligned} F^i_{1\mathrm{SS}} &= \epsilon^6\left[-\frac{15M_1^{jk}M_2^{jl}r^k_{12}r^l_{12}r^i_{12}}{r^7_{12}} + \frac{3M_1^{jk}M_2^{jk}r^i_{12}}{r^5_{12}}\right. \\ &\qquad \left. -\frac{3M_1^{ij}M_2^{jk}r^k_{12}}{r^5_{12}} - \frac{3M_1^{jk}M_2^{ki}r^j_{12}}{r^5_{12}}\right], \\ &= \epsilon^6\frac{1}{r^4_{12}}\left[15(\vec{n}_{12}\cdot\vec{s}_1)(\vec{n}_{12}\cdot\vec{s}_2)n^i_{12} - 3s^i_1(\vec{n}_{12}\cdot\vec{s}_2)\right. \\ &\qquad \left. -3s^i_2(\vec{n}_{12}\cdot\vec{s}_1) - 3n^i_{12}(\vec{s}_1\cdot\vec{s}_2)\right], \end{aligned} \tag{B.14}$$

which is in agreement with, for example, Damour (1982), Kidder (1995), and Thorne and Hartle (1985).

B.0.3 Quadrupole–orbit coupling force

The quadrupole–orbit coupling force can be derived by evaluating the integral

$$F^i_{1\mathrm{QO}} = -\oint_{\partial B_1} dS_j \,{}_8\left[(-g)t^{ij}_{\mathrm{LL}}\right] + \ldots, \tag{B.15}$$

with ${}_8\left[(-g)t^{ij}_{\mathrm{LL}}\right] = \left(\delta^i_k\delta^j_l + \delta^j_k\delta^i_l + -\delta^{ij}\delta_{kl}\right){}_4h^{\tau\tau,k}{}_8h^{\tau\tau,l}/4 + \ldots.$ We then arrive at the quadrupole–orbit coupling force in the same form as that given in Thorne and Hartle (1985), namely,

$$F^i_{1\mathrm{QO}} = \epsilon^4\frac{3}{2r^4_{12}}\left(m_1I_2^{\langle kl\rangle} + m_2I_1^{\langle kl\rangle}\right)\left(2\delta^{il}n^k_{12} - 5n^i_{12}n^k_{12}n^l_{12}\right). \tag{B.16}$$

This result agrees formally with the Newtonian quadrupole–orbit coupling force [compare with (2.56)].

B.0.4 Spin geodesic precession

The spin precession can be evaluated using the following equation:

$$\frac{dM_A^{ij}}{d\tau} = -2\epsilon^{-2} v_A^{[i} P_A^{j]} - 2\epsilon^{-2} R_A^{[ij]}. \tag{B.17}$$

Evaluating the surface integrals in R_A^{ij} and Q_A^i, appearing through the momentum-velocity relation, we have up to ϵ^2,

$$\begin{aligned}
&\frac{d}{d\tau}\left(M_1^{ij} + 2v_1^{[i} D_1^{j]}\right) \\
&= \epsilon^2 \frac{m_2}{r_{12}} \left(2n_{12}^{[i} M_1^{j]k} v_1^k - 4v_1^{[i} M_1^{j]k} n_{12}^k - 4n_{12}^{[i} M_1^{j]k} v_2^k + 4v_2^{[i} M_1^{j]k} n_{12}^k\right) \\
&+\epsilon^2 \frac{6m_2}{r_{12}^3} n_{12}^k n_{12}^{[i} I_1^{j]k} \\
&+\epsilon^2 \frac{2m_2}{r_{12}^3} \left(n_{12}^k n_{12}^i Z_1^{k[jl]l} - n_{12}^k n_{12}^j Z_1^{k[il]l} + n_{12}^k n_{12}^i Z_1^{j[kl]l} - n_{12}^k n_{12}^i Z_1^{j[kl]l}\right) \\
&+O(\epsilon^3).
\end{aligned} \tag{B.18}$$

Note that there is no monopole–monopole coupling. In our formalism the above form is sufficient since we use M_A^{ij} rather than the spin vector s_A^i. To transform the above equation into the usual form we adopt a somewhat unsophisticated method; we treat one star, say the star 1, as though it felt only the gravitational field of the companion star. Motivated by the formalism on extended bodies by Dixon (1979) we introduce the *intrinsic spin 4-vector* $\mathcal{S}_{A\mu}$ and the *intrinsic spin tensor* $\mathcal{M}_A^{\mu\nu}$ by

$$\mathcal{M}_A^{\mu\nu} = \frac{2\epsilon^{-6}}{\sqrt{-g}} \int_{\mathcal{B}_A} d^3\Sigma_\rho\, \sigma_A^{[\mu} \Lambda_N^{\nu]\rho}, \tag{B.19}$$

$$\mathcal{S}_{A\mu} = \frac{1}{2} \epsilon_{\alpha\rho\sigma\mu} \mathcal{M}_A^{\rho\sigma} u_A^\alpha, \tag{B.20}$$

where $\epsilon_{\alpha\rho\sigma\mu}$ is the four-dimensional Levi-Civita permutation symbol with $\epsilon_{0123} = 1$ and $d^3\Sigma_\alpha$ is the proper volume element satisfying $d^3\Sigma_{[\alpha} u_{A\beta]} = 0$. $\mathcal{B}_A$ is a 3-sphere surrounding star A whose normal is $u_{A\alpha}$ and whose radius is ϵR_A. σ_A^μ is the space-like 4-vector satisfying $g_{\mu\nu}\sigma_A^\mu u_A^\mu = 0$. The 4-momentum is normalized as $g_{\mu\nu} u_A^\mu u_A^\nu = \epsilon^{-2}$. The above definitions imply the spin supplementary condition

$$\mathcal{S}_{A\mu} u_A^\mu = 0, \qquad \text{or equivalently} \qquad \mathcal{D}_A^\mu = -\mathcal{M}_A^{\mu\nu} u_{A\mu} = 0, \tag{B.21}$$

where $\mathcal{D}_A^\mu$ is the *intrinsic dipole moment*. Now we construct a coordinate transformation from the near zone $x^\mu = (\tau, x^i)$ to the Fermi normal coordinates $\sigma_A^{\hat{\rho}} = (\hat{\tau}, \sigma^{\hat{i}})$ [see for example Section 40.7 in Misner *et al.* (1973)],

$$\sigma^{\hat{\rho}}_A = e^{\hat{\rho}}_{A\mu} x^\mu, \tag{B.22}$$

$$e^{\hat{\tau}}_{1\tau} = 1 + \frac{1}{2}\epsilon^2 v_1^2 - \epsilon^2 \frac{m_2}{r_{12}}, \tag{B.23}$$

$$e^{\hat{\tau}}_{1i} = -\epsilon^2 v_{1i}, \tag{B.24}$$

$$e^{\hat{i}}_{1j} = \left(1 + \epsilon^2 \frac{m_2}{r_{12}}\right) \delta^i_j + \frac{1}{2}\epsilon^2 v_1^i v_{1j}. \tag{B.25}$$

Then we express the intrinsic spin tensor in the Fermi normal coordinates in terms of the moments in the near zone. Using $d^3\Sigma_\alpha = -u_{A\alpha}\left(1 + \epsilon^2 m_2/\, r_{12} - \epsilon^2 v_1^2/\, 2\right)$ we have

$$\mathcal{M}_1^{\hat{i}\hat{\tau}} = \left[\left(1 + \frac{1}{2}\epsilon^2 v_1^2 - \epsilon^2 \frac{2m_2}{r_{12}}\right)\delta^{ij} - \frac{1}{2}\epsilon^2 v_1^i v_1^j\right] D_1^j - \epsilon^2 M_1^{ik} v_1^k + O(\epsilon^3), \tag{B.26}$$

$$\mathcal{M}_1^{\hat{i}\hat{j}} = M_1^{ij} + \epsilon^2 v_1^k v_1^{[i} M_1^{j]k} + O(\epsilon^3). \tag{B.27}$$

Defining the *intrinsic center of mass* by setting $\mathcal{D}^\mu_A = \mathcal{M}_A^{\hat{i}\hat{\tau}} = 0$ we obtain

$$D^i_A = \epsilon^2 M_1^{ik} v_1^k. \tag{B.28}$$

We note that this relation provides the spin–orbit coupling force in a form previously obtained (see section B.0.1 above, which implies we should take $d^i_A = \mathcal{D}^i_A$ in the present treatment). Using (B.18) we finally arrive at the spin geodesic equation in its usual form, where we omit the quadrupole and Z_1^{ijkl} term for simplicity,

$$\frac{d\vec{\mathcal{S}}_1}{d\tau} = \epsilon^2 \frac{m_2}{r_{12}^2}\left[\left(2\vec{v}_2 - \frac{3}{2}\vec{v}_1\right) \times \vec{n}_{12}\right] \times \vec{\mathcal{S}}_1 + O(\epsilon^3), \tag{B.29}$$

where $\vec{\mathcal{S}}_1$ is the spatial part of the intrinsic spin 4-vector $\mathcal{S}_{1\hat{i}} = \epsilon_{ijk}\mathcal{M}_1^{\hat{j}\hat{k}}/2$. Equation (B.29) is the geodesic precession equation or the so-called de Sitter–Fokker precession equation [see for example Section 40.7 in Misner *et al.* (1973), Brumberg (1991), and Soffel (1989)].

B.0.5 Remarks

In this appendix we have derived the spin–orbit coupling force, the spin–spin coupling force, and the quadrupole–orbit coupling force to the lowest order, along with the spin geodesic equation. Our results agree with the results of others modulo the definition of the center of mass. It should be noted however that we have defined our multipole moments M_A^{ij} and I_A^{ij} as volume integrals of $\Lambda_N^{\mu\nu}$ which manifestly include the effect of the strong internal gravity of the star. Thus our results support the applicability of the spin–orbit, spin–spin, and quadrupole–orbit coupling forces to a relativistic compact binary where the component stars have a strong internal gravitational field. For the effect of these multipole moments on the orbital evolution and/or the gravitational waveform see for example Apostolatos (1995, 1996*a*, 1996*b*), Apostolatos *et al.* (1994), Bildsten and Cutler (1992), Cutler *et al.* (1993), Kidder (1995), Kidder *et al.* (1993), Kochanek (1992), Königsdörffer and Gopakumar (2005), Owen *et al.* (1998), Poisson (1998), and Tagoshi *et al.* (2001).

Appendix C
Null geodesic congruences

Let k^μ be the components of a null vector field, in coordinates z^μ, in a space-time with metric tensor components $g_{\mu\nu}$. Take the integral curves of k^μ to be geodesics with r (say) an affine parameter along them. Then we can write $k^\mu = \partial z^\mu/\partial r$, $g_{\mu\nu}k^\mu\,k^\nu = k_\mu\,k^\mu = 0$ and

$$k^\mu{}_{|\nu}k^\nu = 0, \tag{C.1}$$

with the stroke indicating covariant differentiation with respect to the Riemannian connection associated with the metric tensor $g_{\mu\nu}$. The integral curves of the vector field k^μ constitute a null geodesic congruence. Along any curve C of the congruence let ζ^μ be an infinitesimal connecting vector joining each point of C to a point of any neighboring curve having the same value of the affine parameter r, then ζ^μ is transported along C according to the transport law

$$\zeta^\mu{}_{|\nu}\,k^\nu = k^\mu{}_{|\nu}\,\zeta^\nu. \tag{C.2}$$

Multiplying this by k_μ we see that it implies

$$\frac{\partial}{\partial r}(k_\mu\,\zeta^\mu) = 0, \tag{C.3}$$

so that the scalar product $k_\mu\,\zeta^\mu$ is constant along the null geodesic C. At any point P of C let e^μ be a unit time-like vector. Thus $e_\mu\,e^\mu = 1$. The physical significance of introducing e^μ will emerge below. For the moment we note that we can extend e^μ to a unit time-like vector field along C by parallel transport and thus $e^\mu{}_{|\nu}k^\nu = 0$. We can use e^μ to normalize k^μ by requiring $k_\mu\,e^\mu = 1$. Then we can define a parallel transported, null vector field along C by

$$l^\mu = e^\mu - \frac{1}{2}k^\mu, \tag{C.4}$$

satisfying

$$l^\mu\,l_\mu = 0, \qquad l^\mu\,k_\mu = 1. \tag{C.5}$$

At P on C the pair of null vectors k^μ, l^μ can be completed to a null tetrad by the addition of a complex null vector m^μ and its complex conjugate $\bar{m}^\mu$, and so $m_\mu\,m^\mu = 0 = \bar{m}_\mu\,\bar{m}^\mu$, satisfying

$$m_\mu\,\bar{m}^\mu = -1, \qquad m_\mu\,k^\mu = \bar{m}_\mu\,k^\mu = m_\mu\,l^\mu = \bar{m}_\mu\,l^\mu = 0. \tag{C.6}$$

The vectors $m^\mu, \bar{m}^\mu$ can be extended to vector fields on C by parallel transport. Thus $m^\mu{}_{|\nu}k^\nu = \bar{m}^\mu{}_{|\nu}k^\nu = 0$. The metric tensor can be written in terms of the tetrad as

$$g_{\mu\nu} = -m_\mu\,\bar{m}_\nu - \bar{m}_\mu\,m_\nu + k_\mu\,l_\nu + l_\mu\,k_\nu. \tag{C.7}$$

Returning to the infinitesimal connecting vector ζ^μ we place the following relations upon it:

$$\zeta_\mu\,e^\mu = 0 \qquad \text{and} \qquad (\delta^\mu_\nu - e^\mu\,e_\nu)\,k^\nu\,\zeta_\mu = 0. \tag{C.8}$$

Hence ζ^μ is orthogonal to e^μ and also orthogonal to the projection of k^μ orthogonal to e^μ. We will describe presently a physical interpretation of these relations. Clearly they imply that ζ^μ is orthogonal to e^μ and to k^μ and thus by (C.4) is orthogonal to both l^μ and k^μ. Hence we can write

$$\zeta^\mu = \bar{\zeta}\,m^\mu + \zeta\,\bar{m}^\mu, \tag{C.9}$$

where ζ is a complex-valued function of the coordinates z^μ. Substituting this into (C.2) results in

$$\frac{\partial\bar{\zeta}}{\partial r}\,m^\mu + \frac{\partial\zeta}{\partial r}\,\bar{m}^\mu = \bar{\zeta}\,k^\mu{}_{|\nu}\,m^\nu + \zeta\,k^\mu{}_{|\nu}\,\bar{m}^\nu, \tag{C.10}$$

from which we conclude, on multiplying by m_μ, that

$$\frac{\partial\zeta}{\partial r} = -\sigma\,\bar{\zeta} - \rho\,\zeta, \tag{C.11}$$

where

$$\sigma = k_{\mu|\nu}\,m^\mu\,m^\nu, \qquad \rho = k_{\mu|\nu}\,m^\mu\,\bar{m}^\nu. \tag{C.12}$$

Notwithstanding appearances the *modulus* of the complex variable σ and the real and imaginary parts of the complex variable ρ are independent of the choice of null tetrad. In other words they can be calculated from a knowledge only of $g_{\mu\nu}$ and k^μ. Clearly the *argument* of the complex variable σ *does* depend upon the choice of null tetrad since a simple change of tetrad of the form $m^\mu \to e^{i\psi}m^\mu$, for ψ a real constant, does not leave σ invariant. Expanding the tensor with components $k_{\mu|\nu}$ on the null tetrad, bearing in mind that $k^\mu\,k_{\mu|\nu} = 0 = k^\nu\,k_{\mu|\nu}$, $k_{\mu|\nu}$ is real and the definitions of σ and ρ in (C.12), we find that

$$\begin{aligned} k_{\mu|\nu} &= a\,k_\mu\,k_\nu + b\,k_\mu\,m_\nu + \bar{b}\,k_\mu\,\bar{m}_\nu + c\,m_\mu\,k_\nu + \bar{c}\,\bar{m}_\mu\,k_\nu \\ &\quad + \bar{\sigma}\,m_\mu\,m_\nu + \sigma\,\bar{m}_\mu\,\bar{m}_\nu + \bar{\rho}\,m_\mu\,\bar{m}_\nu + \rho\,\bar{m}_\mu\,m_\nu, \end{aligned} \tag{C.13}$$

where a is a real-valued function of the coordinates and b, c are complex-valued functions of the coordinates, with the bar as always denoting complex conjugation. We find by direct calculation, using the scalar products listed in (C.5) and (C.6), that

$$k^\mu{}_{|\mu} = -\rho - \bar{\rho}, \tag{C.14}$$

$$k_{\mu|\nu}\,k^{\mu|\nu} = 2\,\sigma\,\bar{\sigma} + 2\,\rho\,\bar{\rho}, \tag{C.15}$$

$$k_{\nu|\mu}\,k^{\mu|\nu} = 2\,\sigma\,\bar{\sigma} + \rho^2 + \bar{\rho}^2, \tag{C.16}$$

from which we conclude that $\rho = -\theta - i\omega$ with

$$\theta = \frac{1}{2}k^{\mu}{}_{|\mu} \qquad \text{and} \qquad \omega^2 = \frac{1}{2}k_{[\mu|\nu]}\,k^{\mu|\nu} \geq 0, \tag{C.17}$$

where the square brackets denote skew-symmetrization and

$$|\sigma| = \sqrt{\frac{1}{2}k_{(\mu|\nu)}\,k^{\mu|\nu} - \theta^2}, \tag{C.18}$$

with the round brackets denoting symmetrization. Since $|\sigma|, \theta$, and ω are determined only by the metric and the tangent to the null geodesic congruence it makes sense to look for a physical/geometrical interpretation of these scalars which have the potential to distinguish one null geodesic congruence from another. For this we turn to (C.11) and the relations (C.8) involving the infinitesimal connecting vector ζ^{μ}.

At an event P on the null geodesic C we consider e^{μ} to be the 4-velocity of a small plane circular opaque disk placed in the path of a small bundle of photons with null geodesic world-lines in the neigborhood of C. The first condition involving ζ^{μ} in (C.8) means that ζ^{μ}, the position vector of points on the boundary of the disk, lies in the rest-frame of the disk. The second condition in (C.8) requires the disk to be oriented relative to the photon paths so that the photons strike the disk at right angles when viewed in the rest-frame of the disk. If Q is an event on C an affine parameter distance dr into the future from the event P then at Q we consider e^{μ} to be the 4-velocity of a small plane screen on which a shadow of the disk is projected. The position vector of points on the boundary of the shadow is given by ζ^{μ}, which lies in the rest-frame of the screen on account of the first of (C.8). On account of the second of (C.8) the disk is oriented relative to the photon paths so that the photons strike the screen at right angles when observed in the rest-frame of the screen. In passing from the disk to the screen $r \to r + dr$ and ζ in (C.11) is infinitesimally changed according to

$$\zeta \to \zeta' = (1 - \rho\, dr)\,\zeta - \sigma\,\bar{\zeta}\,dr. \tag{C.19}$$

Since the disk is assumed circular we can take $\zeta = e^{i\phi}$, with $0 \leq \phi \leq 2\pi$, in (C.19). Thus the complex variable ζ' specifies points on the boundary of the shadow relative to points on the boundary of the disk which are specified by ζ. Writing as above $\rho = -\theta - i\omega$ we consider first the case $\omega = 0 = \sigma$. In this case (C.19) reduces to $\zeta' = (1 + \theta\, dr)\,\zeta$ and thus we see that the shadow is a circle whose radius is larger than the disk radius if $\theta > 0$ and smaller than the disk radius if $\theta < 0$. For this reason we interpret the scalar θ as the *expansion* (if positive) or *contraction* (if negative) of the congruence. Next we consider the case $\theta = 0 = \sigma$. Now (C.19) becomes $\zeta' = (1 + i\omega\, dr)\,\zeta = e^{i\omega\, dr}\zeta$, neglecting $O(dr^2)$-terms. Hence in this case the shadow is also a circle of the same radius as the disk but points on the boundary of the shadow are rotated relative to points on the boundary of the disk through a small angle $\omega\, dr$. The scalar ω is referred to as the *twist* of the congruence for this reason. Finally the case $\theta = 0 = \omega$ results in (C.19) becoming $\zeta' = e^{i\phi} - \sigma\, e^{-i\phi}dr$. Hence the shadow is elliptical with the semi-major and semi-minor axes corresponding to ϕ given by $e^{2i\phi} = \pm(\sigma/\bar{\sigma})^{1/2}$. The lengths of these axes are thus $l_{\pm} = 1 \pm |\sigma|\, dr$ and their ratio is therefore $l_+/l_- = 1 + 2\,|\sigma|\, dr$ approximately. The area of the shadow is the same as the area of the circular disk (neglecting $O(dr^2)$-terms) and thus we call the scalar $|\sigma|$ the *shear* of the null geodesic congruence. In practice the complex scalar σ is usually referred to as the *complex shear* of the congruence.

In section 1 of Chapter 5 the generators of the future null cones with vertices on the time-like world-line $r = 0$ in Minkowskian space-time are the null geodesic integral curves of the vector field k^μ. The derivatives of k^μ are given by formula (5.11). From these derivatives one readily verifies that this null geodesic congruence has vanishing twist and shear and has expansion $\theta = (1/2)k^\mu{}_{,\mu} = r^{-1}$. In section 2 of Chapter 5 a twist-free, geodesic null congruence is defined via the 1-form (5.37). This congruence has complex shear σ and expansion θ given by (5.50) and (5.51) respectively. The complex shear is calculated using the first of (C.12) with m^μ and $\bar{m}^\mu$ given via the 1-form $\sqrt{2}\, m_\mu\, dz^\mu = \vartheta^1 + i\vartheta^2$ and its complex conjugate, with ϑ^1 and ϑ^2 given by (5.46) and (5.47).

The theory summarized in this appendix is commonly referred to in the literature as the *Ehlers–Sachs theorem* which first appeared in Ehlers *et al.* (1961). Particularly elegant treatments of it can be found in Pirani (1965) and Chandrasekhar (1983).

Appendix D
Perturbed field equations

Writing, for the perturbed space-time of Chapter 6, $\mathcal{M}^a = \hat{F}^{ab}{}_{|b}$ and $\mathcal{E}_{ab} = \hat{R}_{ab} - 2\,\hat{E}_{ab}$ (as always the tetrad indices $a, b = 1, 2, 3, 4$) the leading terms and the errors we tolerate in these expressions, and thus the extent to which we satisfy the vacuum Einstein–Maxwell field equations, are given here. If the equations of motion are required with greater accuracy then the field equations have to be solved with greater accuracy resulting in smaller errors in the coefficients of these powers of r.

$$\mathcal{M}^A = O_3 \times r^{-2} + O_2 \times r^{-1} + O_1 + O(r),\ (A = 1, 2), \tag{D.1}$$
$$\mathcal{M}^3 = O_3 \times r^{-1} + O_1 + O(r), \tag{D.2}$$
$$\mathcal{M}^4 = O_2 \times r + O(r^2), \tag{D.3}$$

and

$$\mathcal{E}_{AA} = O_3 \times r^{-4} + O_3 \times r^{-2} + O_2 \times r^{-1} + O(r^0), \tag{D.4}$$
$$\mathcal{E}_{11} - \mathcal{E}_{22} = O_3 \times r^{-2} + O_2 \times r^{-1} + O_1 + O(r), \tag{D.5}$$
$$\mathcal{E}_{12} = O_3 \times r^{-2} + O_2 \times r^{-1} + O_1 + O(r), \tag{D.6}$$
$$\mathcal{E}_{A3} = O_3 \times r^{-2} + O_1 \times r^{-1} + O_1 + O(r), \tag{D.7}$$
$$\mathcal{E}_{A4} = O_2 \times r^{-3} + O_2 \times r^{-2} + O_2 \times r^{-1} + O(r^0), \tag{D.8}$$
$$\mathcal{E}_{33} = O_1 + O(r), \tag{D.9}$$
$$\mathcal{E}_{34} = O_4 \times r^{-4} + O_3 \times r^{-3} + O_2 \times r^{-2} + O_2 \times r^{-1} + O(r^0), \tag{D.10}$$
$$\mathcal{E}_{44} = O_3 \times r^{-3} + O_3 \times r^{-2} + O(r^{-1}). \tag{D.11}$$

The coefficients of the various powers of r in these expressions are calculated by substituting into $\mathcal{M}^a$ and $\mathcal{E}_{ab}$ the metric tensor, given via the line-element (5.45) with the expansions (6.1)–(6.6), and the potential 1-form (5.63), with the expansions (6.7)–(6.9). Writing (D.1)–(D.11) in the form

$$\mathcal{M}^a = \sum {}_{(n)}\mathcal{M}^a\, r^n, \qquad \mathcal{E}_{ab} = \sum {}_{(n)}\mathcal{E}_{ab}\, r^n, \tag{D.12}$$

the coefficients ${}_{(n)}\mathcal{M}^a$ and ${}_{(n)}\mathcal{E}_{ab}$ required to establish (D.1)–(D.11) are given below. The coefficients required from Maxwell's equations with index 1 or 2 are

$$
\begin{aligned}
{}_{(-2)}\mathcal{M}^1 &= \frac{\partial}{\partial y}\left\{\hat{P}_0^2\left(\frac{\partial \hat{M}_0}{\partial x}-\frac{\partial \hat{L}_0}{\partial y}\right)\right\}+2e^2\hat{L}_2-e\,\hat{a}_{-1}\,\hat{P}_0^{-2}+\frac{\partial \hat{K}_{-1}}{\partial x}\\
&\quad +O_3, &\text{(D.13)}\\
{}_{(-1)}\mathcal{M}^1 &= O_3, &\text{(D.14)}\\
{}_{(0)}\mathcal{M}^1 &= 6\,m\,L_3+O_2=O_1, &\text{(D.15)}\\
{}_{(-2)}\mathcal{M}^2 &= \frac{\partial}{\partial x}\left\{\hat{P}_0^2\left(\frac{\partial \hat{M}_0}{\partial x}-\frac{\partial \hat{L}_0}{\partial y}\right)\right\}-2e^2\hat{M}_2+e\,\hat{b}_{-1}\hat{P}_0^{-2}-\frac{\partial \hat{K}_{-1}}{\partial y}\\
&\quad +O_3, &\text{(D.16)}\\
{}_{(-1)}\mathcal{M}^2 &= O_3, &\text{(D.17)}\\
{}_{(0)}\mathcal{M}^2 &= -6\,m\,M_3+O_2=O_1, &\text{(D.18)}
\end{aligned}
$$

and the remaining required coefficients, with index 3 or 4, are

$$
\begin{aligned}
{}_{(-1)}\mathcal{M}^3 &= O_2, &\text{(D.19)}\\
{}_{(0)}\mathcal{M}^3 &= -4mP_0^{-2}F_{\mu\nu}k^\mu\,v^\nu+O_2=O_1, &\text{(D.20)}\\
{}_{(1)}\mathcal{M}^4 &= -2\hat{P}_0^{-2}\hat{K}_1+4\hat{P}_0^{-2}\,(\hat{a}_{-1}\hat{L}_2+\hat{b}_{-1}\hat{M}_2-e\,\hat{q}_2)+2\left(\frac{\partial \hat{L}_2}{\partial x}+\frac{\partial \hat{M}_2}{\partial y}\right)\\
&\qquad +O_2. &\text{(D.21)}
\end{aligned}
$$

Writing

$$
\hat{\Delta}=\hat{P}_0^2\left(\frac{\partial^2}{\partial x^2}+\frac{\partial^2}{\partial y^2}\right), \tag{D.22}
$$

we find that the required coefficients from Einstein's equations involving indices 1, 2 are given by

$$
\begin{aligned}
{}_{(-4)}\mathcal{E}_{AA} &= O_3, &\text{(D.23)}\\
{}_{(-3)}\mathcal{E}_{AA} &= O_3, &\text{(D.24)}
\end{aligned}
$$

$$
\begin{aligned}
{}_{(-2)}\mathcal{E}_{AA} = {}& 2\,\hat{\Delta}\log\hat{P}_0 - 4\,\hat{P}_0^4\left(\frac{\partial\hat{M}_0}{\partial x} - \frac{\partial\hat{L}_0}{\partial y}\right)\left(\frac{\partial\hat{M}_2}{\partial x} - \frac{\partial\hat{L}_2}{\partial y}\right)\\
& -3\hat{P}_0^2\left(\frac{\partial}{\partial x}\left(\hat{P}_0^{-2}\hat{a}_{-1}\right) + \frac{\partial}{\partial y}\left(\hat{P}_0^{-2}\hat{b}_{-1}\right)\right)\\
& +4\,e\hat{K}_1 + 4\,\hat{K}_1\,\hat{K}_{-1} - 8\,e(\hat{a}_{-1}\hat{L}_2 + \hat{b}_{-1}\hat{M}_2)\\
& -2\,\hat{c}_0 + 12\,e^2\hat{q}_2 - \frac{1}{2}\hat{P}_0^{-2}\left(\hat{a}_{-1}^2 + \hat{b}_{-1}^2\right) + O_3,
\end{aligned}
\tag{D.25}
$$

$$
\begin{aligned}
{}_{(-1)}\mathcal{E}_{AA} = {}& -4\,\hat{c}_1 - 8\,\frac{\partial\log\hat{P}_0}{\partial u} + 8\,e\,\hat{K}_2 - 32\,m\,\hat{q}_2\\
& -4\,\hat{P}_0^2\left(\frac{\partial}{\partial x}\left(\hat{P}_0^{-2}\hat{a}_0\right) + \frac{\partial}{\partial y}\left(\hat{P}_0^{-2}\hat{b}_0\right)\right) + O_2,
\end{aligned}
\tag{D.26}
$$

$$
\begin{aligned}
{}_{(-2)}\mathcal{E}_{11} - {}_{(-2)}\mathcal{E}_{22} = {}& \frac{\partial\hat{b}_{-1}}{\partial y} - \frac{\partial\hat{a}_{-1}}{\partial x} - 4\,e^2\hat{\alpha}_2 - \frac{1}{2}\hat{P}_0^{-2}\left(\hat{a}_{-1}^2 + \hat{b}_{-1}^2\right)\\
& -8\,e^2\hat{P}_0^2\left(\hat{L}_2^2 - \hat{M}_2^2\right) - 8\,\hat{P}_0^2\hat{L}_2\,\frac{\partial\hat{K}_{-1}}{\partial x} + 8\,\hat{P}_0^2\hat{M}_2\,\frac{\partial\hat{K}_{-1}}{\partial y}\\
& +O_3,
\end{aligned}
\tag{D.27}
$$

$$
\begin{aligned}
{}_{(-1)}\mathcal{E}_{11} - {}_{(-1)}\mathcal{E}_{22} = {}& 16\,m\,\hat{P}_0^2\left(\hat{L}_2^2 - \hat{M}_2^2\right) - 8\,\hat{P}_0^2\hat{L}_2\,\frac{\partial\hat{K}_0}{\partial x} + 8\,\hat{P}_0^2\hat{M}_2\,\frac{\partial\hat{K}_0}{\partial y}\\
& +16\,m\,\hat{\alpha}_2 - 2\,\frac{\partial\hat{a}_0}{\partial x} + 2\,\frac{\partial\hat{b}_0}{\partial y} + O_2,
\end{aligned}
\tag{D.28}
$$

$$
\begin{aligned}
{}_{(0)}\mathcal{E}_{11} - {}_{(0)}\mathcal{E}_{22} = {}& -8\hat{c}_0\,\hat{P}_0^2\left(\hat{L}_2^2 - \hat{M}_2^2\right) - 8\,\hat{P}_0^2\hat{L}_2\,\frac{\partial\hat{K}_1}{\partial x} + 8\,\hat{P}_0^2\hat{M}_2\,\frac{\partial\hat{K}_1}{\partial y}\\
& -12\,\hat{\alpha}_2\,\hat{c}_0 - 3\frac{\partial\hat{a}_1}{\partial x} + 3\frac{\partial\hat{b}_1}{\partial y} + O_1 = O_1,
\end{aligned}
\tag{D.29}
$$

$$
\begin{aligned}
{}_{(-2)}\mathcal{E}_{12} = {}& -\frac{1}{2}\frac{\partial\hat{a}_{-1}}{\partial y} - \frac{1}{2}\frac{\partial\hat{b}_{-1}}{\partial x} - 4\,e^2\hat{\beta}_2 - \frac{1}{2}\hat{P}_0^{-2}\hat{a}_{-1}\,\hat{b}_{-1}\\
& -8\,e^2\hat{P}_0^2\hat{L}_2\hat{M}_2 - 4\,\hat{P}_0^2\hat{L}_2\,\frac{\partial\hat{K}_{-1}}{\partial y} - 4\,\hat{P}_0^2\hat{M}_2\,\frac{\partial\hat{K}_{-1}}{\partial x}\\
& +O_3,
\end{aligned}
\tag{D.30}
$$

$$
\begin{aligned}
{}_{(-1)}\mathcal{E}_{12} = {}& 16\,m\,\hat{P}_0^2\hat{L}_2\,\hat{M}_2 - 4\,\hat{P}_0^2\hat{L}_2\,\frac{\partial\hat{K}_0}{\partial y} - 4\,\hat{P}_0^2\hat{M}_2\,\frac{\partial\hat{K}_0}{\partial x}\\
& +8\,m\,\hat{\beta}_2 - \frac{\partial\hat{a}_0}{\partial y} - \frac{\partial\hat{b}_0}{\partial x} + O_2,
\end{aligned}
\tag{D.31}
$$

$$
{}_{(0)}\mathcal{E}_{12} = -8\hat{c}_0\,\hat{P}_0^2\hat{L}_2\,\hat{M}_2 - 4\,\hat{P}_0^2\hat{M}_2\,\frac{\partial\hat{K}_1}{\partial x} - 4\,\hat{P}_0^2\hat{L}_2\,\frac{\partial\hat{K}_1}{\partial y}
$$

$$
-6\,\hat{\beta}_2\,\hat{c}_0 - \frac{3}{2}\frac{\partial\hat{a}_1}{\partial y} - \frac{3}{2}\frac{\partial\hat{b}_1}{\partial x} + O_1 = O_1. \tag{D.32}
$$

The remaining coefficients required from Einstein's equations are

$$
{}_{(-2)}\mathcal{E}_{13} = 4\,\hat{M}_2\,\hat{P}_0^3\left(\frac{\partial\hat{L}_0}{\partial y} - \frac{\partial\hat{M}_0}{\partial x}\right) + 4\,e\,\hat{L}_2\hat{P}_0 + \hat{P}_0^{-1}\hat{a}_{-1}
$$

$$
+4\,\hat{K}_{-1}\,\hat{L}_2\,\hat{P}_0 + O_3, \tag{D.33}
$$

$$
{}_{(-1)}\mathcal{E}_{13} = 6\,e\,P_0\,L_3 + O_2 = O_1, \tag{D.34}
$$

$$
{}_{(0)}\mathcal{E}_{13} = -4\,\hat{K}_1\,\hat{L}_2\,\hat{P}_0 + 2\,\hat{P}_0^3\left(\frac{\partial}{\partial x}\left(\hat{P}_0^{-2}\hat{\alpha}_2\right) + \frac{\partial}{\partial y}\left(\hat{P}_0^{-2}\hat{\beta}_2\right)\right)
$$

$$
+2\,\hat{P}_0\,\frac{\partial\hat{q}_2}{\partial x} - 2\,\hat{P}_0^{-1}\hat{a}_1 + 4\,\hat{P}_0^3\hat{M}_2\left(\frac{\partial\hat{L}_2}{\partial y} - \frac{\partial\hat{M}_2}{\partial x}\right)
$$

$$
+O_1, \tag{D.35}
$$

$$
{}_{(-3)}\mathcal{E}_{14} = 4\,m\,e\,P_0\,F_{\mu\nu}\frac{\partial k^\mu}{\partial x}\,v^\nu + O_3 = O_2, \tag{D.36}
$$

$$
{}_{(-2)}\mathcal{E}_{14} = 2\,e\,\hat{c}_0\hat{L}_2\,\hat{P}_0 + 2\,e\,\hat{P}_0\frac{\partial\hat{K}_1}{\partial x} + \frac{1}{2}\,\hat{P}_0^{-1}\hat{a}_{-1}\,\hat{c}_0
$$

$$
-\frac{1}{2}\hat{P}_0\frac{\partial}{\partial y}\left(\hat{P}_0^2\left\{\frac{\partial}{\partial y}\left(\hat{P}_0^{-2}\hat{a}_{-1}\right) - \frac{\partial}{\partial x}\left(\hat{P}_0^{-2}\hat{b}_{-1}\right)\right\}\right)
$$

$$
-\hat{P}_0^{-1}\hat{a}_{-1}\,\hat{\Delta}\log\hat{P}_0 + O_2 = O_2, \tag{D.37}
$$

$$
{}_{(-1)}\mathcal{E}_{14} = O_2, \tag{D.38}
$$

$$
{}_{(-2)}\mathcal{E}_{23} = -4\,\hat{L}_2\,\hat{P}_0^3\left(\frac{\partial\hat{L}_0}{\partial y} - \frac{\partial\hat{M}_0}{\partial x}\right) + 4\,e\,\hat{M}_2\hat{P}_0 + \hat{P}_0^{-1}\hat{b}_{-1}
$$

$$
+4\,\hat{K}_{-1}\,\hat{M}_2\,\hat{P}_0 + O_3, \tag{D.39}
$$

$$
{}_{(-1)}\mathcal{E}_{23} = 6\,e\,P_0\,M_3 + O_2 = O_1, \tag{D.40}
$$

$$
{}_{(0)}\mathcal{E}_{23} = -4\,\hat{K}_1\,\hat{M}_2\,\hat{P}_0 + 2\,\hat{P}_0^3\left(-\frac{\partial}{\partial y}\left(\hat{P}_0^{-2}\hat{\alpha}_2\right) + \frac{\partial}{\partial x}\left(\hat{P}_0^{-2}\hat{\beta}_2\right)\right)
$$

$$
+2\,\hat{P}_0\,\frac{\partial\hat{q}_2}{\partial y} - 2\,\hat{P}_0^{-1}\hat{b}_1 - 4\,\hat{P}_0^3\hat{L}_2\left(\frac{\partial\hat{L}_2}{\partial y} - \frac{\partial\hat{M}_2}{\partial x}\right)
$$

$$
+O_1, \tag{D.41}
$$

$$ {}_{(-3)}\mathcal{E}_{24} = 4\,m\,e\,P_0\,F_{\mu\nu}\frac{\partial k^\mu}{\partial y}\,v^\nu + O_3 = O_2, \tag{D.42} $$

$$ \begin{aligned} {}_{(-2)}\mathcal{E}_{24} &= 2\,e\,\hat{c}_0\hat{M}_2\,\hat{P}_0 + 2\,e\,\hat{P}_0\frac{\partial \hat{K}_1}{\partial y} + \frac{1}{2}\,\hat{P}_0^{-1}\hat{b}_{-1}\,\hat{c}_0 \\ &\quad + \frac{1}{2}\hat{P}_0\,\frac{\partial}{\partial x}\left(\hat{P}_0^2\frac{\partial}{\partial y}\left(\hat{P}_0^{-2}\hat{a}_{-1}\right) - \hat{P}_0^2\frac{\partial}{\partial x}\left(\hat{P}_0^{-2}\hat{b}_{-1}\right)\right) \\ &\quad - \hat{P}_0^{-1}\hat{b}_{-1}\,\hat{\Delta}\log\hat{P}_0 + O_2 = O_2, \end{aligned} \tag{D.43} $$

$$ {}_{(-1)}\mathcal{E}_{24} = O_2, \tag{D.44} $$

and

$$ {}_{(0)}\mathcal{E}_{33} = -8\,\hat{P}_0^2\left(\hat{L}_2^2 + \hat{M}_2^2\right) + 12\,\hat{q}_2 + O_3, \tag{D.45} $$

$$ {}_{(-4)}\mathcal{E}_{34} = O_4, \tag{D.46} $$

$$ {}_{(-3)}\mathcal{E}_{34} = O_3, \tag{D.47} $$

$$ \begin{aligned} {}_{(-2)}\mathcal{E}_{34} &= 2\,e\,\hat{K}_1 + \frac{1}{2}\hat{P}_0^2\left(\frac{\partial}{\partial x}\left(\hat{P}_0^{-2}\hat{a}_{-1}\right) + \frac{\partial}{\partial y}\left(\hat{P}_0^{-2}\hat{b}_{-1}\right)\right) \\ &\quad + O_2, \end{aligned} \tag{D.48} $$

$$ \begin{aligned} {}_{(-1)}\mathcal{E}_{34} &- \hat{c}_1 + 2\,\hat{P}_0^{-1}\frac{\partial \hat{P}_0}{\partial u} + 4\,e\,\hat{K}_2 + 8\,m\,\hat{q}_2 \\ &\quad + \hat{P}_0^2\left(\frac{\partial}{\partial x}\left(\hat{P}_0^{-2}\hat{a}_0\right) + \frac{\partial}{\partial y}\left(\hat{P}_0^{-2}\hat{b}_0\right)\right) + O_2, \end{aligned} \tag{D.49} $$

$$ {}_{(-4)}\mathcal{E}_{44} = O_3, \tag{D.50} $$

$$ \begin{aligned} {}_{(-3)}\mathcal{E}_{44} &= 2\,m\,\hat{P}_0^2\left(\frac{\partial}{\partial x}\left(\hat{P}_0^{-2}\hat{a}_{-1}\right) + \frac{\partial}{\partial y}\left(\hat{P}_0^{-2}\hat{b}_{-1}\right)\right) \\ &\quad - 4\,e^2\hat{P}_0\,\frac{\partial \hat{P}_0}{\partial u} - 2\,\hat{\Delta}\hat{f}_{-1} + O_3, \end{aligned} \tag{D.51} $$

$$ {}_{(-2)}\mathcal{E}_{44} = T_1 + T_2 + T_3 + O_3. \tag{D.52} $$

In the final equation here the terms T_1, T_2, T_3 are given in the text by (6.54), (6.56), and (6.58) respectively.

References

Abramovici, A., Althouse, W. E., Drever, R. W. P., Gursel, Y., Kawamura, S., Raab, F. J., Shoemaker, D. H., Sievers, L., Spero, R. E., Thorne, K. S., Vogt, R. E., Weiss, R., Whitcomb, S. E., and Zucker, M. E. (1992). *Science*, **256**, 325–333.

Anderson, J. L. and DeCanio, T. C. (1975). *Gen. Rel. and Grav.*, **6**, 197–237.

Antonacopoulos, G. and Tsoupakis, E. (1979). *Astrophys. and Space Sci.*, **62**, 185–188.

Apostolatos, T. A. (1995). *Phys. Rev. D*, **52**, 605–620.

Apostolatos, T. A. (1996*a*). *Phys. Rev. D*, **54**, 2421–2437.

Apostolatos, T. A. (1996*b*). *Phys. Rev. D*, **54**, 2438–2441.

Apostolatos, T. A., Cutler, C., Sussman, G. J., and Thorne, K. S. (1994). *Phys. Rev. D*, **49**, 6274–6297.

Arzoumanian, Z., Cordes, J. M., and Wasserman, I. (1999). *Astrophys. J.Phys.*, **520**, 696–705.

Asada, H. and Futamase, T. (1997*a*). *Phys. Rev. D*, **56**, 6062–6066.

Asada, H. and Futamase, T. (1997*b*). *Prog. Theor. Phys. Suppl.*, **128**, 123–181.

Ashby, N. and Bertotti, B. (1986). *Phys. Rev. D*, **34**, 2246–2258.

Baker, J. G., Centrella, J., Choi, D. I., Koppitz, M., and vanMeter, J. (2006). *Phys. Rev. Lett.*, **96** (111102).

Barker, B. M. and O'Connell, R. F. (1975*a*). *Astrophys. J. Lett.*, **199**, L25–L26.

Barker, B. M. and O'Connell, R. F. (1975*b*). *Phys. Rev. D*, **12**, 329–335.

Bergmann, P. G. (1947). *An Introduction to the Theory of Relativity.* Prentice Hall, New York.

Berti, E., Iyer, S., and Will, C. M. (2006). *Phys. Rev. D*, **74** (061503(R)).

Bildsten, L. and Cutler, C. (1992). *Astrophys. J.*, **400**, 175–180.

Blanchet, L. (2006). *Living Rev. Relativity*, **9** (4).

Blanchet, L. and Damour, T. (1988). *Phys. Rev. D*, **37**, 1410–1435.

Blanchet, L. and Damour, T. (1989). *Ann. Inst. Henri Poincaré A*, **50**, 377–408.

Blanchet, L., Damour, T., Esposito-Farèse, G., and Iyer, B. R. (2004). *Phys. Rev. D*, **69** (124007).

Blanchet, L. and Faye, G. (2000*a*). *J. Math. Phys.*, **41**, 7675–7714.

Blanchet, L. and Faye, G. (2000*b*). *Phys. Lett. A*, **271**, 58–64.

Blanchet, L. and Faye, G. (2001*a*). *J. Math. Phys.*, **42**, 4391–4418.

Blanchet, L. and Faye, G. (2001*b*). *Phys. Rev. D*, **63** (062005).

Blanchet, L., Faye, G., and Ponsot, B. (1998). *Phys. Rev. D*, **58** (124002).

Blanchet, L. and Schäfer, G. (1989). *Mon. Not. R. Astron. Soc.*, **239**, 845–867.

Blandford, R. and Teukolsky, S. A. (1975). *Astrophys. J. Lett.*, **198**, L27–L29.

Blandford, R. and Teukolsky, S. A. (1976*a*). *Astrophys. J.*, **205**, 580–591.

Blandford, R. and Teukolsky, S. A. (1976*b*). *Astrophys. J.*, **205**, 580–591.

Bradaschia, C., Fabbro, R. Del, Virgilio, A. Di, Giazotto, A., Kaurzky, H., Montelatici, V., Passuello, D., Brillet, A., Cregut, O., Hello, P., Man, C. N., Manh, P. T., Marraud, A., Shoemaker, D. H., Vinet, J. Y., Barone, F., di Fiore, L., Milano, L., Russo, G., Aguirregabiria, J. M., Bel, H., Duruisseau, J. P., Denmat, G. Le, Tourrenc, P., Capozzi, M., Longo, M., Lops, M., Pinto, I., Rotoli, G., Damour, T., Bonazzola, S., Marck, J. A., Gourghoulon, Y., Holloway, L. E., Fuligni, F., Iafolla, V., and Natale, G. (1990). *Nucl. Instrum. Methods A*, **289**, 518–525.

Breton, R. P., Kaspi, V. M., Kramer, M., McLaughlin, M. A., Lyutikov, M., Ransom, S. M., Stairs, I. H., Ferdman, R. D., Camilo, F., and Possenti, A. (2008). *Science*, **321**, 104–107.

Brill, D. R. and Hartle, J. B. (1964). *Phys. Rev.*, **135**, B271–B278.

Brumberg, V. A. (1991). *Essential Relativistic Celestial Mechanics.* Adam Hilger, Bristol.

Brumberg, V. A. and Kopeikin, S. M. (1989). *Nuovo Cimento B*, **103**, 63–98.

Burgay, M., D'Amico, N., Possenti, A., Manchester, R. N., Lyne, A. G., Joshi, B. C., Laughlin, M. A. Mc, Kramer, M., Sarkissian, J. M., and et al., F. Camilo (2003). *Nature*, **426**, 531–533.

Burke, W. L. (1971). *J. Math. Phys.*, **12**, 401–418.

Campanelli, M., Dettwyler, M., Hannam, M., and Lousto, C. O. (2006*a*). *Phys. Rev. D*, **74** (087503).

Campanelli, M., Lousto, C. O., Marronetti, P., and Zlochower, Y. (2006*b*). *Phys. Rev. Lett.*, **96** (111101).

Centrella, J. (2003). *The Astrophysics of Gravitational Wave Sources.* Maryland, pp. 249–373. American Institute of Physics.

Chandrasekhar, S. (1965). *Astrophys. J.*, **142**, 1488–1540.

Chandrasekhar, S. (1969). *Astrophys. J.*, **158**, 45–54.

Chandrasekhar, S. (1983). *The Mathematical Theory of Black Holes.* Oxford University Press, Oxford.

Chandrasekhar, S. and Esposito, F. P. (1970). *Astrophys. J.*, **160**, 153–179.

Chandrasekhar, S. and Nutku, Y. (1969). *Astrophys. J.*, **158**, 55–79.

Chenciner, A. and Montgomery, R. M. (2000). *Ann. Math.*, **152**, 881–901.

Chiba, T., Imai, T., and Asada, H. (2007). *Mon. Not. R. Astron. Soc.*, **377**, 269–272.

Curran, S. J. and Lorimer, D. R. (1995). *Mon. Not. R. Astron. Soc.*, **276**, 347–352.

Cutler, C., Apostolatos, T. A., Bildsten, L., Finn, L. S., Flanagan, É., Kennefick, D., Markovic, D. M., Ori, A., Poisson, E., and Sussman, G. J. (1993). *Phys. Rev. Lett.*, **70**, 2984–2987.

Cutler, C. and Thorne, K. S. (2002). *General Relativity and Gravitation.* Singapore, pp. 72–111. World Scientific.

Damour, T (1982). *C. R. Acad. Sci. Ser.II*, **294**, 1355–1357.

Damour, T. (1983). *Gravitational Radiation.* Amsterdam, pp. 59–144. Elsevier.

Damour, T. (1987*a*). *Gravitation in Astrophysics*: Cargese 1986. New York, pp. 3–62. Plenum Press.

Damour, T. (1987*b*). *Three Hundred Years of Gravitation.* Cambridge, pp. 128–198. Cambridge University Press.

Damour, T. and Deruelle, N. (1981*a*). *Phys. Lett. A*, **87**, 81–84.

Damour, T. and Deruelle, N. (1981*b*). *C. R. Acad. Sci. Ser.II*, **293**, 537–540.
Damour, T. and Deruelle, N. (1981*c*). *C. R. Acad. Sci. Ser.II*, **293**, 877–880.
Damour, T., Jaranowski, P., and Schäfer, G. (2000). *Phys. Rev. D*, **62** (021501).
Damour, T., Jaranowski, P., and Schäfer, G. (2001*a*). *Phys. Lett.*, **513**, 147–155.
Damour, T., Jaranowski, P., and Schäfer, G. (2001*b*). *Phys. Rev. D*, **63** (044021).
Damour, T. and Schäfer, G. (1988). *Nuovo Cimento B*, **101**, 127–176.
Damour, T., Soffel, M., and Xu, C. (1991). *Phys. Rev. D*, **43**, 3273–3307.
Damour, T., Soffel, M., and Xu, C. (1992). *Phys. Rev. D*, **45**, 1017–1044.
Damour, T., Soffel, M., and Xu, C. (1993). *Phys. Rev. D*, **47**, 3124–3135.
Damour, T., Soffel, M., and Xu, C. (1994). *Phys. Rev. D*, **49**, 618–635.
Damour, T. and Taylor, J. H. (1991). *Astrophys. J.*, **366**, 501–511.
Danby, J. M. A. (1988). *Fundamentals of Celestial Mechanics.* William-Bell, Virginia.
Danzmann, K. (1992). *Relativistic Gravity Research with Emphasis on Experiments and Observations.* Berlin, pp. 184–209. Springer.
Daucourt, G. (1997). *Class. and Quantum Grav.*, **14**, A109–A118.
de Andrade, V. C., Blanchet, L., and Faye, G. (2001). *Class. and Quantum Grav.*, **18**, 753–778.
D'Eath, P. D. (1975*a*). *Phys. Rev. D*, **11**, 1387–1403.
D'Eath, P. D. (1975*b*). *Phys. Rev. D*, **12**, 2183–2199.
Detweiler, S. L. (1978). *Astrophys. J.*, **225**, 687–693.
DeWitt, B. S. and Brehme, R. W. (1960). *Ann. Phys. N.Y.*, **9**, 220–259.
Diener, P., Herrmann, F., Pollnay, D., Schnetter, E., Seidel, E., Takahashi, R., Thornburg, J., and Ventrella, J. (2006). *Phys. Rev. Lett.*, **96** (121101).
Dirac, P. A. M. (1938). *Proc. R. Soc. A*, **167**, 148–169.
Dixon, W. G. (1970*a*). *Proc. Roy. Soc. A*, **314**, 499–527.
Dixon, W. G. (1970*b*). *Proc. Roy. Soc. A*, **319**, 509–547.
Dixon, W. G. (1973). *Gen. Rel. and Grav.*, **4**, 199–209.
Dixon, W. G. (1974). *Phil. Trans. Roy. Soc. A*, **277**, 59–119.
Dixon, W. G. (1979). *Isolated Gravitating Systems in General Relativity.* Amsterdam, pp. 156–219. North-Holland.
Ehlers, J. (1997). *Class. and Quantum Grav.*, **14**, A119–A126.
Ehlers, J., Rosenblum, A., Goldberg, J. N., and Havas, P. (1976). *Astrophys. J. Lett.*, **208**, L77–L81.
Ehlers, J., Sachs, R. K., and Jordan, P. (1961). *Akad. Wiss. Mainz Abh., Math.-Naturw.*, **1**, 3–62.
Einstein, A., Infeld, L., and Hoffmann, B. (1938). *Ann. Math.*, **39**, 65–100.
Epstein, R. (1977*a*). *Astrophys. J.*, **216**, 92–100.
Epstein, R. (1977*b*). *Astrophys. J.*, **216**, 92–100.
Epstein, R. (1979). *Astrophys. J.*, **231**, 644–644.
Faye, G., Jaranowski, P., and Schäfer, G. (2004). *Phys. Rev. D*, **69** (124029).
Fermi, E. (1922). *Atti Accad. Naz. Lincei Rend. Cl. Sci. Fis. Mat.Nat.*, **31**, 21, 51.
Finn, L. S. (1996). *Phys. Rev. D*, **53**, 2878–2894.
Fock, V. (1939). *J. Phys. USSR*, **1**, 81–106.
Fock, V. A. (1959). *Theory of Space, Time and Gravitation.* Pergamon Press, London.

Fujiwara, T. and Montgomery, R. (2005). *Pacific J. Math.*, **219**, 271–283.
Futamase, T. (1983). *Phys. Rev. D*, **28**, 2373–2381.
Futamase, T. (1985*a*). *Phys. Rev. D*, **32**, 2566–2574.
Futamase, T. (1985*b*). *Phys. Rev. D*, **32**, 2566–2574.
Futamase, T. (1987). *Phys. Rev. D*, **36**, 321–329.
Futamase, T. and Hogan, P. A. (1993). *J. Math. Phys.*, **34**, 154–169.
Futamase, T., Hogan, P. A., and Itoh, Y. (2008). *Phys. Rev. D*, **78** (104014).
Futamase, T. and Itoh, Y. (2007). *Living Rev. Relativity*, **10**(2).
Futamase, T. and Schutz, B. F. (1983). *Phys. Rev. D*, **28**, 2363–2372.
Galan, J., Munoz-Almaraz, F. J., Freire, E., Doedel, E., and Vanderbauwhede, A. (2002). *Phys. Rev. Lett.*, **88** (241101).
Geroch, R. (1969). *Comm. Math. Phys.*, **13**, 180–193.
Gultekin, K., Miller, M. C., and Hamilton, D. P. (2006). *Astrophys. J.*, **640**, 156–166.
Hadamard, J. (1932). *Le Problèm de Cauchy et les Équations aux Dérivées Partielles Linéaries Hyperbolique*. Hermann, Paris.
Heggie, D. C. (2000). *Mon. Not. R. Astron. Soc.*, **318**, L61–L63.
Hobbs, J. M. (1968). *Ann. Phys. N.Y.*, **47**, 141–165.
Hogan, P. A. and Ellis, G. F. R. (1989). *Ann. Phys. N.Y.*, **195**, 293–323.
Hogan, P. A. and Imaeda, M. (1979). *J. Phys, A: Math. Gen.*, **12**, 1061–1069.
Hogan, P. A. and McCrea, J. D. (1974). *Gen. Rel. and Grav.*, **5**, 79–113.
Hogan, P. A. and Robinson, I. (1985). *Found. Phys.*, **15**, 617–627.
Hogan, P. A. and Trautman, A. (1987). *Gravitation and Geometry*. Naples, pp. 215–242. Bibliopolis.
Hulse, R. A. and Taylor, J. H. (1975). *Astrophys. J. Lett.*, **195**, L51–L53.
Imai, T., Chiba, T., and Asada, H. (2007). *Phys. Rev. Lett.*, **98** (201102).
Infeld, L. and Plebanski, J. (1960). *Motion and Relativity*. Pergamon Press, London.
Ioka, K., Chiba, T., Tanaka, T., and Nakamura, T. (1998). *Phys. Rev. D*, **58** (063003).
Isaacson, R. A. (1968*a*). *Phys. Rev.*, **166**, 1263–1271.
Isaacson, R. A. (1968*b*). *Phys. Rev.*, **166**, 1271–1280.
Isaacson, R. A., Welling, J. S., and Winicour, J. (1984). *Phys. Rev. Lett.*, **53**, 1870–1872.
Itoh, Y. (2004). *Phys. Rev. D*, **69** (064018).
Itoh, Y. (2010). *gr-qc Archive* (0911.4232).
Itoh, Y. and Futamase, T. (2003). *Phys. Rev. D*, **68** (121501(R)).
Itoh, Y., Futamase, T., and Asada, H. (2000). *Phys. Rev. D*, **62** (064002).
Itoh, Y., Futamase, T., and Asada, H. (2001). *Phys. Rev. D*, **63** (064038).
Jaranowski, P. and Schäfer, G. (1998*a*). *Phys. Rev. D*, **57**, 7274–7291.
Jaranowski, P. and Schäfer, G. (1998*b*). *Phys. Rev. D*, **57**, 5948–5949.
Jaranowski, P. and Schäfer, G. (1999). *Phys. Rev. D*, **60** (124003).
Jaranowski, P. and Schäfer, G. (2000). *Ann. Phys. (Berlin)*, **9**, 378–383.
Jaranowski, P. and Schäfer, G. (2001). *Phys. Rev. D*, **63** (029902).
Kalogera, V., Kim, C., Lorimer, D. R., Burgay, M., D'Amico, N., Possenti, A., Manchester, R. N., Lyne, A. G., Joshi, B. C., McLaughlin, M. A., Kramer, M., Sarkissian, J. M., and Camilo, F. (2004). *Astrophys. J. Lett.*, **601**, L179–L182.

Kalogera, V., Narayan, R., Spergel, D. N., and Taylor, J. H. (2001). *Astrophys. J.*, **556**, 340–356.
Kapela, T. and Zgliczynski, P. (2003). *Nonlinearity*, **16**, 1899–1918.
Kates, R. E. (1980*a*). *Phys. Rev. D*, **22**, 1871–1878.
Kates, R. E. (1980*b*). *Phys. Rev. D*, **22**, 1853–1870.
Kerlick, G. D. (1980*a*). *Gen. Rel. and Grav.*, **12**, 467–482.
Kerlick, G. D. (1980*b*). *Gen. Rel. and Grav.*, **12**, 521–543.
Kidder, L. E. (1995). *Phys. Rev. D*, **52**, 821–847.
Kidder, L. E., Will, C. M., and Wiseman, A. G. (1993). *Phys. Rev. D*, **47**, R4183–R4187.
Kim, C., Kalogera, V., and Lorimer, D. R. (2003). *Astrophys. J.*, **584**, 985–995.
Klioner, S. A. (2003). *Astron. J.*, **125**, 1580–1597.
Klioner, S. A. and Kopeikin, S. M. (1994). *Astrophys. J.*, **427**, 951–955.
Kochanek, C. S. (1992). *Astrophys. J.*, **398**, 234–247.
Kokkotas, K. and Schmidt, B. (1999). *Living Rev. Relativity*, **2**(2).
Königsdörffer, C. and Gopakumar, A. (2005). *Phys. Rev. D*, **71** (024039).
Kopeikin, S. M. (1985). *Sov. Astron.*, **29**, 516–.
Kopeikin, S. M. and Schäfer, G. (1999). *Phys. Rev. D*, **60** (124002).
Kuroda, K. (1997). *Gravitational Waves: Sources and Detectors*. Singapore. World Scientific.
Landau, L. D. and Lifshitz, E. M. (1975). *The Classical Theory of Fields* (4th. edn). Pergamon Press, London.
Levi-Civita, T. (1965). *The N-Body Problem in General Relativity* (2nd edn). Reidel, Dordrecht.
Marchal, C. (1990). *The Three Body Problem*. Elsevier, Amsterdam.
Markovic, D. (1993). *Phys. Rev. D*, **48**, 4738–4756.
McCrea, J. D. (1981). *Gen. Rel. and Grav.*, **13**, 397–415.
McCrea, J. D. and O'Brien, G (1978). *Gen. Rel. and Grav.*, **9**, 1101–1118.
Mio, N. (2006*a*). *Selected Papers from the Sixth Edoardo Amaldi Conference on Gravitational Waves*. Institute of Physics, Bristol.
Mio, N. (2006*b*). *Sixth Edoardo Amaldi Conference on Gravitational Waves*. Institute of Physics, Bristol.
Misner, C. W., Thorne, K. S., and Wheeler, J. A. (1973). *Gravitation*. W. H. Freeman, New York.
Molenda, T. (1984). Contributed Papers: Proceedings of 10th. International Conference on General Relativity and Gravitation 1983. Dordrecht, p. 97. Reidel.
Montgomery, R. (2005). *Ergodic Theory and Dynamical Systems*, **25**, 921–947.
Moore, C. (1993). *Phys. Rev. Lett.*, **70**, 3675–3679.
Nakamura, T. and Oohara, K. (1983). *Phys. Lett. A*, **98**, 403–406.
Nakamura, T. and Sasaki, M. (1981). *Phys. Lett. B*, **106**, 69–72.
Newman, E. T. and Posadas, R. (1969). *Phys. Rev.*, **187**, 1784–1791.
Newman, E. T. and Unti, T. W. J. (1962). *J. Math. Phys.*, **3**, 891–900.
O'Brien, G. (1979). *Gen. Rel. and Grav.*, **10**, 129–148.
Ohta, T. and Kimura, T. (1988). *Phys. Lett. A*, **129**, 436–439.
Ohta, T. and Kimura, T. (1989*a*). *Prog. Theor. Phys.*, **81**, 679–689.

Ohta, T. and Kimura, T. (1989*b*). *Prog. Theor. Phys.*, **81**, 662–678.
Ohta, T., Okamura, H., Kimura, T., and Hiida, K. (1973). *Prog. Theor. Phys.*, **50**, 492–514.
O'Murchadha, N. and York, J. W. (1974). *Phys. Rev. D*, **10**, 2345–2357.
O'Raifeartaigh, L. (1958). *Proc. R. Irish Acad. A*, **59**, 15–24.
O'Raifeartaigh, L. (1962). *Proc. R. Irish Acad. A*, **62**, 63–72.
Owen, B. J., Tagoshi, H., and Ohashi, A. (1998). *Phys. Rev. D*, **57**, 6168–6175.
Papapetrou, A. (1951*a*). *Proc. Phys. Soc. A*, **64**, 57–75.
Papapetrou, A. (1951*b*). *Proc. Roy. Soc. A*, **209**, 248–258.
Pati, M. E. and Will, C. M. (2000). *Phys. Rev. D*, **62** (124015).
Pati, M. E. and Will, C. M. (2002). *Phys. Rev. D*, **65** (104008).
Peters, P. C. (1964). *Phys. Rev.*, **136**, B1224–B1232.
Peters, P. C. and Mathews, J. (1963). *Phys. Rev.*, **131**, 435–440.
Pirani, F. A. E. (1965). Brandeis Lectures 1964. New Jersey, pp. 249–373. Prentice Hall.
Poisson, E. (1998). *Phys. Rev. D*, **57**, 5287–5290.
Pretorius, F. (2005). *Phys. Rev. Lett.*, **95** (121101).
Price, R. H. and Pullin, J. (1994). *Phys. Rev. Lett.*, **72**, 3297–3300.
Rendall, A. D. (1992). *Proc. R. Soc. A*, **438**, 341–360.
Robinson, I. and Robinson, J. R. (1972). General Relativity: Papers in Honour of J. L. Synge. Oxford, pp. 151–166. Clarendon Press.
Robinson, I. and Trautman, A. (1960). *Phys. Rev. Lett.*, **4**, 431–432.
Robinson, I. and Trautman, A. (1962). *Proc. R. Soc. A*, **270**, 463–473.
Roy, A. E. (1988). *Orbital Motion.* Institute of Physics, Bristol.
Sachs, R. K. (1962). *Proc. R. Soc. A*, **270**, 103–126.
Sasaki, M. and Tagoshi, H. (2003). *Living Rev. Relativity*, **6**(6).
Schäfer, G. (1985*a*). *Ann. Phys. N.Y.*, **161**, 81–100.
Schäfer, G. (1986). *Gen. Rel.* and *Grav.*, **18**, 255–270.
Schutz, B. F. (1980). *Phys. Rev. D*, **22**, 249–259.
Schutz, B. F. (1984). *Relativistic Astrophysics and Cosmology.* Singapore, p. 35. World Scientific.
Schutz, B. F. (1986). *Nature*, **323**, 310–311.
Schutz, B. F. (2002). *Lighthouses of the Universe: The Most Luminous Celestial Objects and Their Use for Cosmology.* Springer, Berlin.
Schwartz, L. (1966). *Théorie des Distributions.* Hermann, Paris.
Shibata, M. (1999). *Phys. Rev. D*, **60** (104052).
Simo, C. (2002). *Contemp. Math.*, **292**, 209–228.
Smarr, L. L. (1979). *Sources of Gravitational Radiation.* Cambridge. Cambridge University Press.
Sobolev, S. L. (1989). *Partial Differential Equations of Mathematical Physics.* Dover, New York.
Soffel, M. H. (1989). *Relativity in Astrometry, Celestial Mechanics and Geodesy.* Springer, Berlin.
Stergioulas, N. (2003). *Living Rev. Relativity*, **6**(3).

Stewart, J. M. and Walker, M. (1974). *Proc. R. Soc. A*, **341**, 49–74.

Synge, J. L. (1970*a*). *Proc. R. Irish Acad. A*, **69**, 11–38.

Synge, J. L. (1970*b*). *Annali di Matematica Pura ed Applicata*, **84**, 33–60.

Tagoshi, H., Ohashi, A., and Owen, B. J. (2001). *Phys. Rev. D*, **63** (044006).

Teukolsky, S. A. (1973). *Astrophys. J.*, **185**, 635–647.

Thorne, K.S. (1980). *Rev. Mod. Phys.*, **52**, 299–339.

Thorne, K. S. (1994). *Relativistic Cosmology*. Tokyo. Universal Academy Press.

Thorne, K. S. and Hartle, J. B. (1985). *Phys. Rev. D*, **31**, 1815–1837.

VandenHeuvel, E. P. J. and Lorimer, D. R. (1996). *Mon. Not. R. Astron. Soc.*, **283**, L37–L39.

Wald, R. M. (1984). *General Relativity*. University of Chicago Press, Chicago.

Walker, M. and Will, C. M. (1980*a*). *Astrophys. J. Lett.*, **242**, L129–L133.

Walker, M. and Will, C. M. (1980*b*). *Phys. Rev. Lett.*, **45**, 1741–1744.

Wang, T., Stebbins, A., and Turner, E. L. (1996). *Phys. Rev. Lett.*, **77**, 2875–2878.

Wardell, Z. E. (2002). *Mon. Not. R. Astron. Soc.*, **334**, 149–155.

Weisberg, J. M. and Taylor, J. H. (2005). *The Relativistic Binary Pulsar* B1913+16: *Thirty Years of Observations and Analysis*. San Francisco, pp. 25–32. Astronomical Society of the Pacific.

Wheeler, J. C. (1975). *Astrophys. J. Lett.*, **196**, L67–L70.

Will, C. M. (1987). *Three Hundred Years of Gravitation*. Cambridge, pp. 80–127. Cambridge University Press.

Will, C. M. (1993). *Theory and Experiment in Gravitational Physics* (2nd. edn). Cambridge University Press, Cambridge.

Will, C. M. (1994). *Relativistic Cosmology*. Tokyo. Universal Academy Press.

Will, C. M. (1999). *Prog. Theor. Phys. Suppl.*, **136**, 158–167.

Will, C. M. (2005). *Phys. Rev. D*, **71** (084027).

Will, C. M. (2006). *Living Rev. Relativity*, **9**(3).

Will, C. M. and Wiseman, A. G. (1996). *Phys. Rev. D*, **54**, 4813–4848.

Zwart, S. P., Baumgardt, H., Hut, P., Makino, J., and McMillan, S. (2004). *Nature*, **428**, 724–726.

Index

The manufacturer's authorised representative in the EU for product safety is Oxford University Press España S.A. of El Parque Empresarial San Fernando de Henares, Avenida de Castilla, 2 - 28830 Madrid (www.oup.es/en or product.safety@oup.com). OUP España S.A. also acts as importer into Spain of products made by the manufacturer.
Printed and bound by CPI Group (UK) Ltd, Croydon, CR0 4YY
14/09/2026
02216295-0007